49 Electronic 12-Volt Projects

Delton T. Horn

FIRST EDITION
FIRST PRINTING

Library of Congress Cataloging-in-Publication Data

Horn, Delton T.
49 electronic 12-volt projects/ by Delton T. Horn.
p. cm.
ISBN 0-8306-9265-7 ISBN 0-8306-3265-4 (pbk.)
1. Electronics—Amateurs' manuals. I. Title. II. Title: Forty-nine electronic twelve-volt projects.
TK9965.H637 1990
621.381'078—dc20 89-27497
CIP

TAB BOOKS Inc. offers software for sale. For information and a catalog, please contact TAB Software Department, Blue Ridge Summit, PA 17294-0850.

Questions regarding the content of this book should be addressed to:

Reader Inquiry Branch
TAB BOOKS Inc.
Blue Ridge Summit, PA 17294-0214

Acquisions Editor: Roland S. Phelps
Technical Editor: Steven L. Burwen
Production: Katherine Brown

Contents

List of Projects

Introduction

This book features almost fifty projects for the electronics experimenter to build and enjoy. Some have serious applications, while others are just for fun.

An unusual feature of this book is that the projects are all designed to operate from +12 volts dc. This can be very convenient for the experimenter. You don't have to bother with building a new power supply for each new project. You don't have to fool around with several different battery sizes. If you use a variable power supply, you will be less likely to accidentally hook up a circuit with the power supply set for the wrong voltage.

Some prior experience with electronics project building is assumed in this book, but most of the projects are simple enough for the beginner who knows which end of a soldering iron to hold.

In selecting the projects to include in this book, I tried to offer as much variety as possible. The 49 projects include LED flashers, amplifiers, digital circuits, tone generators, timers, counters, and many other useful and interesting electronic devices. Several of the projects are intended for automotive use. These projects can be powered from a car's 12-volt battery. An easy way to power a project in an automobile is to use a plug that fits into the cigarette lighter.

I encourage you to experiment with the circuits shown here and to modify them to suit your own individual applications. The choice of projects was made according to three criteria:

- Is the project fun?
- Is the project useful?
- Can the hobbyist learn from this project?

I had fun creating the projects for this book. I hope you enjoy building and experimenting with them. If you enjoy this book, you might also be interested in the companion volume, *49 Electronic 6-Volt Projects* (TAB Book #3275).

❖1 Getting Started

THE BULK OF THIS BOOK CONTAINS OVER FOUR-DOZEN ELECTRONIC projects. Each project is designed to operate from 12-volts dc. Many other project books can be frustrating, especially when projects call for unusual and sometimes difficult-to-obtain supply voltages.

This first chapter is a basic introduction to electronics project building for the beginner. More experienced hobbyists might also want to read this introductory chapter as a convenient review. In addition to the basics of project constructions, this chapter also covers the various power-supply options you can use to run your projects.

BREADBOARDING

You could just build the projects directly, constructing them on simple perf boards with point-to-point wiring or on simple home-brew printed-circuit boards. Before you reach for your soldering iron, however, I strongly suggest that you try to breadboard each of the projects first. This is a good idea for almost any electronics project. It is much easier to troubleshoot a problem or make modifications to a circuit if you don't have to waste a lot of time desoldering and resoldering connection points.

In addition to the disadvantages of being time consuming and a real nuisance, resoldering can introduce a number of new problems. Solder bridges (creating short circuits between adjacent leads or connection points) become increasingly likely every time you apply the soldering iron. Reheated solder (during resoldering) is especially prone to flowing someplace it shouldn't go. In addition, excessive heat can very easily damage or destroy some components, especially semiconductors, which are inherently thermal sensitive and somewhat delicate. Obviously, it is best to apply heat as infrequently (and for as short a time) as possible.

Breadboard the circuits first. When you are 100 percent sure that everything is right and that the circuit does exactly what you want it to do, then solder a permanent version of the project. It might seem like an unnecessary extra step, but in the long run, breadboarding can really save you a lot of time, frustration, and even money.

Using a solderless socket (Fig. 1-1) to breadboard a circuit also encourages experimentation—which is unquestionably the best way to learn about electronics. In some cases, you might not need or want a permanent version of certain projects. You might only want to experiment with a circuit for awhile. In such cases, a solderless breadboard is the only reasonable way to go. There will be a minimum of fuss and bother, and you can easily reuse the components over and over.

Fig. 1-1 Use a solderless socket to experiment with electronic circuits.

You can use a solderless socket by itself to breadboard your circuits, but this method leaves something to be desired. For one thing, a lot of extra work can be involved. Many projects require external circuits. Typical examples are power supplies (discussed later in this chapter) and oscillators (to serve as signal sources). It is a nuisance to have to build such a basic, common circuit from scratch every time you want to breadboard a project.

A second disadvantage of the solderless socket is that it makes it awkward at best to work with large components (such as switches, potentiometers, transformers, and the like). These large components cannot be mounted in the small holes of the solderless socket.

The solution to both these problems is to use a full breadboarding system. Such a system is built around a solderless socket, but it is more than just the socket itself.

One (or sometimes more than one) solderless socket is mounted on a secure base. The base is actually a housing for several common circuits such as power supplies, oscillators, amplifiers, etc. that might be needed frequently when breadboarding projects. Convenient connection points are provided to run wires from these circuits to the breadboarded circuit being built upon the solderless socket.

Several common, large components also are mounted in the case. These components might include one or more switches, one or more potentiometers, a speaker, and possibly a transformer or an LED (or LCD) display unit.

You can buy ready-made breadboarding systems from a number of manufacturers, or you could design and build a customized breadboarding system of your own. Coming up with your own system is an especially good idea if you do a lot of work with a particular type of project and have special requirements.

There are some precautions you should be aware of when breadboarding circuits on a solderless socket (whether you use a full breadboarding system or not). For one thing, component leads generally are exposed. You must take care to avoid short circuits from adjacent wires accidentally touching one another. In some circuits, the exposed component leads also can present an electrical shock hazard if you get careless.

Always remember to make all changes in any breadboarded circuit with the power off to minimize the risk of damaging some (usually expensive) components, especially ICs and other delicate semiconductors. Turning the power off before working on the breadboarded circuit also will reduce significantly the chances of your receiving an electrical shock and possibly suffering severe injury. Never forget that electrical shocks can be very painful or even fatal; always use caution. Reasonable precautions are never a waste of time.

You also should be aware that some circuits might not work well, and in some cases they might not work at all, in a standard, solderless breadboarding socket. Such problems are encountered most often with circuits designed for high-frequency signals. In high-frequency circuits, the length of connecting wires and shielding (or lack thereof) can be of crucial importance. In a breadboarded circuit, "phantom" components (stray capacitances and inductances) also can be a problem. Such "phantom" components generally have very small values, and thus have a negligible effect in most circuits. In high-frequency circuits, however, they can alter considerably the way the circuit works.

Fortunately, the circuit that cannot be prototyped in a breadboarding socket, at least to some degree, is very much the exception than the rule. If high frequencies do happen to be involved in a project you are working on, you can minimize potential problems by keeping the interconnecting wires as short as possible. If the breadboarded circuit operates incorrectly or erratically, try relocating some of the components. This often (though not always) clears up the problem.

Breadboarded circuits sometimes change their operating parameters noticeably when you convert the project to a more permanent form of construction. The changes can be slight, or they can be considerable. You need to be aware of such potential problems when experimenting with a breadboarded circuit. There really isn't very much you can do about it in advance, but at least you can be prepared to recognize such problems when they do crop up. If nothing else, you might be able to save some time and frustrated hair pulling.

Once you have breadboarded and experimented with your project and have gotten all of the bugs out, you will probably want to rebuild some circuits in a more permanent way. Solderless breadboarding sockets are great for testing and experimenting with prototype circuits, but they really aren't much good when it comes to putting the project to practical use.

By definition, breadboarded circuits use temporary connections. In actual use, some component leads can bend easily and touch one another, causing potentially harmful shorts. Components can even fall out of the socket altogether when the device is moved about. Interference signals can be generated easily and picked up by the exposed wiring.

Generally speaking, packaging a circuit built on a solderless socket is tricky at best. They don't tend to fit standard circuit housings and boxes very well. Also, a solderless socket is fairly expensive. It is certainly more than worth the price if it is reused for many different circuits. If you tie it up with a single permanent project, however, you are only cheating yourself. Less expensive construction methods that are more reliable, more compact, and that offer better overall performance are readily available. I briefly discuss some of the more common permanent construction methods in the next few pages.

PERF BOARD

You can construct many relatively simple circuits, including most of the projects presented in this book, on a perforated circuit board, or *perf board*. Figure 1-2 shows a typical perf board. You mount the component leads through the perforated holes in the nonconductive board. The holes are spaced evenly in rows, and most standard components can fit onto such a board. You solder the component leads and jumper wires directly together, using point-to-point wiring.

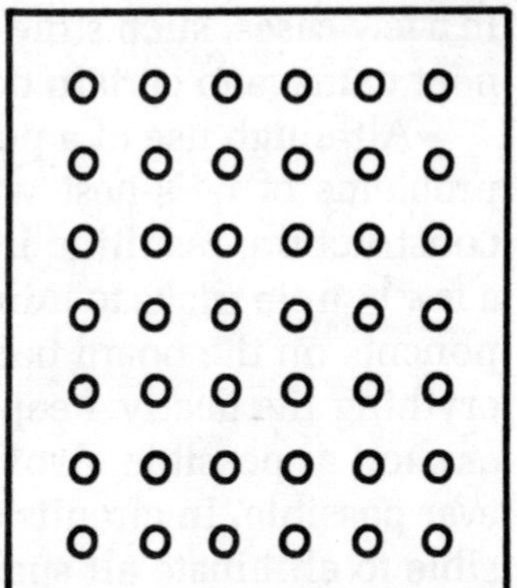

Fig. 1-2 *Many electronic circuits can be built on a "perf board."*

Many hobbyists use flea clips. You insert the flea clip through one of the holes in the perf board, and component leads are attached to the flea clip, rather than directly through the hole itself (see Fig. 1-3).

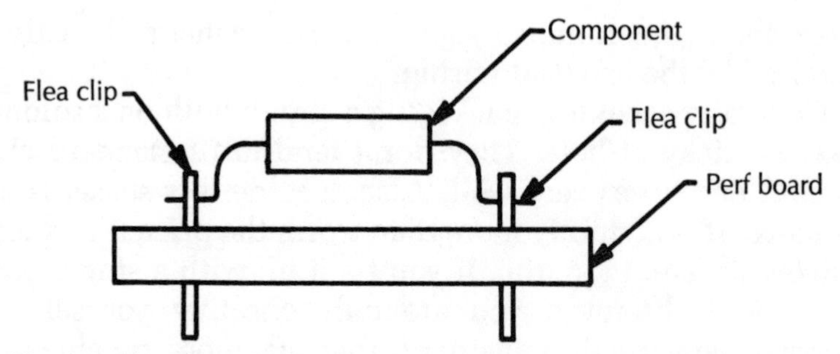

Fig. 1-3 Flea clips often are used to mount components on perf boards.

The perf board serves as a secure base for the mounted circuitry. Only very small, very simple circuits should ever be wired together directly without any supporting circuit board. The board provides physical support and helps prevent a "rat's nest" of jumbled and tangled wiring that is almost impossible to trace if an error is made in construction, or to troubleshoot if the project requires servicing at some later date. Rat's-nest wiring also is an open invitation to short circuits and/or breaks in connecting wires. In addition, a lot of loose hanging wires can create various electrical problems, such as stray capacitances and inductances between nearby wires. Stray capacitances or inductances allow signals to get into the wrong portions of the circuit. Obviously, this will cause erratic operation, if not complete circuit failure. In a few cases, such stray signals could conceivably cause permanent damage to certain components.

Although use of a perf board automatically minimizes some problems of rat's-nest wiring, it is still possible to use sloppy construction, resulting in another rat's nest of sorts. You can use a few simple tricks to minimize such problems. Arrange the components on the board before you begin soldering. Make sure everything fits neatly. Keep all connecting leads and jumper wires as short as possible. Avoid the use of crossed jumper wires whenever possible. In circuits of any complexity, it is probably impossible to eliminate all such wire crossings, but try to arrange the components physically so as few wire crossings as possible are necessary. Of course, any wires that cross one another must be insulated adequately to prevent short circuits. Use straight-line paths for jumper wires whenever possible.

PRINTED CIRCUITS

For moderate to complex circuits, or for circuits from which a number of duplicates will be built, a printed-circuit board gives very good results. Of course, a printed circuit could be designed and used for a simple, one-shot project, but that would be technological overkill. It would probably be more trouble than it's really worth.

On a printed-circuit board, copper traces on one side (or in very complex circuits, on both sides) of the board act as connecting wires between the various components. Very steady, stable, and sturdy connections can be made because the component leads are soldered directly to the supporting board itself.

The original board is a slab of insulating material (like that used in perf boards), with one side (or sometimes both sides) covered with a thin layer of copper. During the preparation process, the desired pattern of connecting traces is applied to the copper-clad side(s) of the board. The pattern can be drawn on directly with special resist ink, or stick-on resist labels can be used. More advanced experimenters and professional manufacturers usually apply the trace pattern with photographic techniques.

Once the pattern of desired copper traces is applied (by whatever means), the board is soaked in a bath of a special acid, or *etchant*. The acid eats away the exposed copper, but the resist ink or labels protect certain portions of the copper cladding. When the board is removed from the etchant bath and washed to remove the excess acid and the resist ink, the only remaining copper is in the form of the desired pattern of traces. In the final stage, holes are drilled to accommodate the leads of the components to be mounted on the board.

You must take great care in laying out a printed circuit (PC) board. You must work out the exact sizes and positions of all components before you apply the resist to the board. If you make a mistake, you'll probably have to do the entire board over. You must take special care to avoid wire crossings. The copper traces cannot cross over each other (unless a dual-sided board is used) because the traces can exist only in two dimensions. If a crossing is absolutely essential, you must use a wire jumper to connect two separated traces. Mount wire jumpers onto PC boards just like ordinary components.

Stray capacitances between traces can adversely affect circuit performance in some projects (especially when high-frequency signals are involved). In critical circuits, a guard band between traces can help reduce the potential problem of stray capacitances. Figure 1-4 illustrates an example of the use of guard bands.

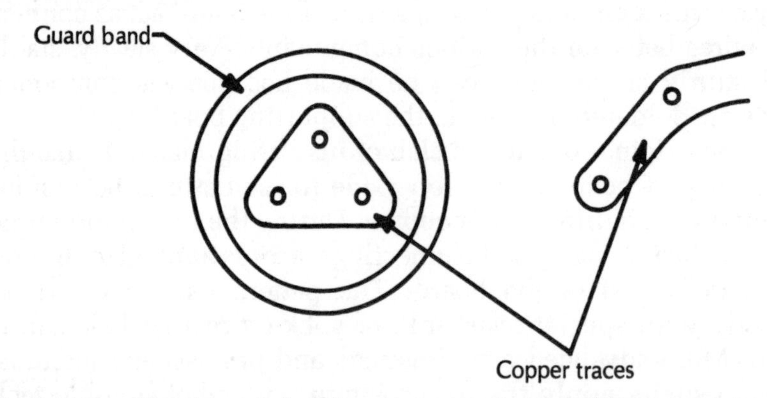

Fig. 1-4 *Guard bands can help reduce stray capacitance problems on PC boards.*

The copper traces on most PC boards usually are placed very close to one another, especially in circuits using ICs, because the IC's leads don't have much spacing between them. Because adjacent traces are close to each other, it is very easy to create a short circuit on a PC board. A small speck of solder or a piece of a component's excess lead could very easily bridge across two (or more) adjacent traces, creating a short. When soldering component leads to a PC board, use a minimum of solder. If you use too much solder, it will tend to flow and bridge across adjacent traces.

Be careful not to apply too much heat when soldering on a printed-circuit board. Don't leave the soldering iron at any one point for too long a time. Excessive heat damages the bond between the copper cladding and the supporting board. The copper trace tends to peel up and away from the board. It then becomes extremely fragile, and breakage is then almost inevitable.

Tiny, nearly invisible cracks in the copper traces also can be problematic if you're not careful at all stages in PC construction.

There are many ways such cracks can be caused. If you use reasonable care, you shouldn't have many problems of this type, but occasionally they crop up, even for the most careful of us. Fairly wide traces that are widely spaced are generally the easiest to work with. Unfortunately, this isn't always practical with all circuits. Often it will be quite impossible, especially where integrated circuits are used.

Printed-circuit-board construction produces very short component leads. Lengthy leads aren't needed. This feature can help minimize interference and stray capacitance problems.

UNIVERSAL PRINTED-CIRCUIT BOARDS

Laying out and etching a PC board is admittedly a lot of fuss and bother. Many hobbyists don't want to take the trouble.

Recently, a new choice has appeared on the list of project construction methods. This is the use of a universal printed-circuit board.

A universal PC board is a commercially available board with a standardized pattern pre-etched onto it. You can use the pattern on the universal PC board to build a great many different projects.

A universal PC board is sort of a cross between a perf board and a dedicated PC board. Like a perf board, it has a number of rows of holes. You place the components where it is convenient (as long as the leads go through to the desired trace). Most of the holes in the board are left unused. An ordinary PC board usually has no excess holes. You drill the holes in precise locations for the specific components used in the circuit.

Several universal PC board designs are available, including designs for analog, digital, and op-amp (analog with two power-supply buses) circuits. You can select whatever pattern is most convenient for your individual project. Radio Shack stores, along with a number of other dealers, carry several universal PC boards.

WIRE-WRAPPING

Today many IC circuits are constructed using a method known as *wire-wrapping*. A thin wire (typically 30 gauge) is wrapped tightly around a special square post. The squared off edges of the post bite into the wire, making good electrical and mechanical connections, without soldering. Components fit into special sockets that connect their leads to the square wrapping posts. Wire-wrapping sockets are available to accommodate most IC DIP packages.

The wire-wrapping construction technique is most appropriate for circuits that are made up primarily of several integrated circuits. If the circuit contains just a few discrete components, they can be fitted into special sockets, or *headers*. Alternatively, you can solder the discrete components directly, while the connections to the ICs are wire-wrapped. This is known as *hybrid construction*. In circuits involving many discrete components, the wire-wrapping method tends to be rather impractical. In these cases, you should use a different method of circuit construction.

Wire-wrapped connections can be made (or unmade) quickly and easily without any risk of potential heat damage to delicate semiconductor components (ICs). This is because no heat source (soldering iron) is applied to the component leads.

There are, however, some disadvantages to using this method to construct your projects. As I mentioned above, discrete components (resistors, capacitors, transistors, etc.) are awkward at best. Also, the thin wire-wrapping wire is very fragile and quite easily broken. Because the wire is so thin, it can be used only to carry very low-power signals. Stray rf pickup and emission can be a problem, especially with high-frequency circuits. The wiring in a wire-wrapped circuit can be very difficult to trace.

Still, when a large number of ICs are involved (some circuits require several dozen), wire-wrapping can be a very convenient way to construct a project.

I describe wire-wrapping here merely for the sake of completeness. It probably would not be a particularly good choice for constructing any of the projects in this book.

SUBSTITUTING COMPONENTS

With very few exceptions (which I always indicate in the text), the components used in these projects are not overly crucial. I have tried to use only readily available components. If you can't get ahold of the exact component called for, you should be able to find a reasonable substitute without much trouble. If you already have something in your junk box that will do the job, there is certainly no point in running out and buying a brand new part with a slightly different identification number.

Use a good substitution guide to find equivalent devices for any semiconductors, especially transistors. ICs tend to be more difficult to find suitable substitutes for. I have tried to use only commonly available devices, but this is a constantly changing field. It is entirely possible that a chip that was widely available when I wrote this book could be discontinued by the time you read this. Your best bet in such cases is to try the various surplus houses. Many advertise in the backs of the electronics hobbyist magazines. Also check your local yellow pages. There might be a surplus house or a well-stocked jobber or supplier near you.

All digital ICs used are the CMOS type. Do not substitute TTL devices. They cannot handle a 12-volt supply voltage.

All resistors are assumed generally to be standard 5 percent, 1/4-watt carbon units, unless otherwise noted. While I have not tested this, 10 percent resistors should work fine in most of the circuits presented here. In most cases, you can substitute values other than those specified in the parts lists. Use a resistor with a value as close as possible to the specified value, except in cases where I encourage experimentation. In some cases, you might want to experiment with substituting a potentiometer (or trimpot) for a fixed resistor.

If you have a problem finding a resistor with a certain desired value (such problems are most likely for oddball values, of course), you can combine two or more available resistors to come up with the hard-to-find value needed in your circuit.

You can string resistors together in series to create a larger effective resistance (Fig. 1-5). To calculate the total effective

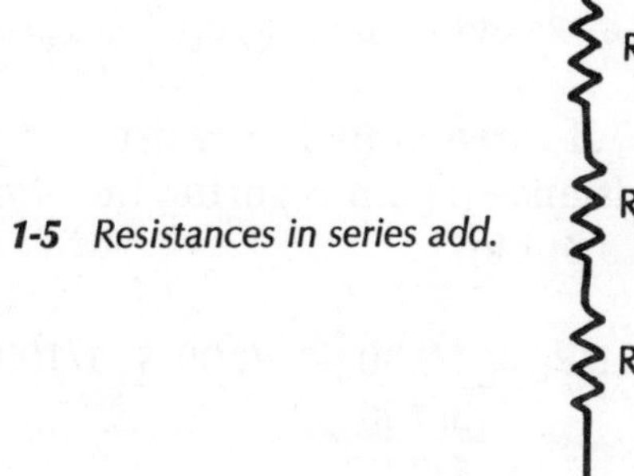

Fig. 1-5 *Resistances in series add.*

resistance, simply add together the individual series values:

$$R_t = R1 + R2 + R3 \ldots + R_n$$

As an example, let's say you have three 180-ohm resistors wired in series. In this case, the total resistance (ignoring the resistor tolerances) is equal to:

$$R_t = 180 + 180 + 180$$
$$= 540 \text{ ohms}$$

Note that the total effective resistance for any series combination of resistors is always larger than any of the individual series resistances.

You also can combine resistors in parallel (Fig. 1-6). The formula for the total effective resistance is slightly more complex for parallel resistances. The reciprocal of the total effective resistance is equal to the sum of the reciprocals of the individual parallel resistances. This sounds very complicated, but it's a lot simpler if you see the actual equation:

$$1/R_t = 1/R1 + 1/R2 + 1/R3 \ldots + 1/R_n$$

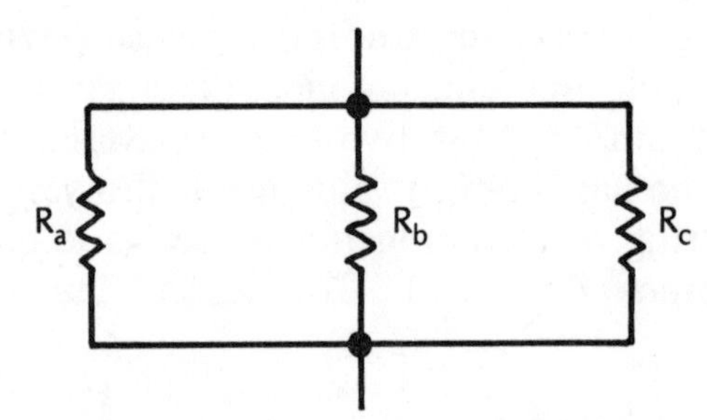

Fig. 1-6 *Resistors also can be combined in parallel.*

For example, if three 100-ohm resistors are in parallel, the total effective resistance (again ignoring the effects of the resistor tolerances), works out to:

$$1/R_t = 1/100 + 1/100 + 1/100$$
$$= 3/100$$
$$= 1/33.3$$
$$= 33.3 \text{ ohms}$$

Note that the total effective resistance of a parallel resistor combination is always less than the value of any of the individual parallel resistances.

If you combine just two resistances in parallel, you can use a slightly different alternate formula if you prefer:

$$R_t = (R1 \times R2)/(R1 + R2)$$

As an example, let's assume R1 is 100 ohms, and R2 has a resistance of 220 ohms. In this case, the total effective resistance (ignoring resistor tolerances) is:

$$\begin{aligned} R_t &= (100 \times 220)/(100 + 220) \\ &= 22{,}000/320 \\ &= 68.75 \text{ ohms} \end{aligned}$$

This alternate formula gives exactly the same results as the general parallel resistance formula, but it is sometimes a bit more convenient to use.

You need to consider one special case. If both parallel resistances have equal values, the total effective resistance is always equal to exactly one-half the value of either of the parallel resistances. For example, if

$$R1 = R2 = 100 \text{ ohms}$$

then the total effective resistance is 50 ohms. You can confirm this by working out the problem with one of our parallel resistance formulas:

$$\begin{aligned} R_t &= (100 \times 100)/(100 + 100) \\ &= 10000/200 \\ &= 100/2 \\ &= 50 \text{ ohms} \end{aligned}$$

Of course, you can use both series and parallel resistance combinations together. Figure 1-7 shows a sample series/parallel network. To determine the total effective resistance of such a resistance network, you need to break the circuit down into individual series or parallel combinations and solve for the total effective resistance step by step.

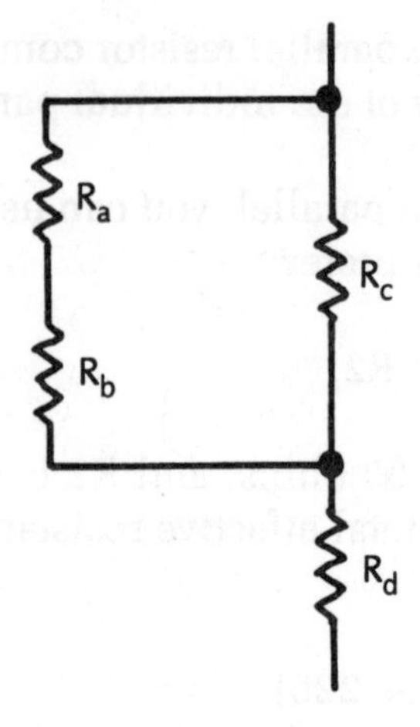

Fig. 1-7 Some circuits use both series and parallel resistances.

In our sample circuit, you should first solve for the series value of resistances R_a and R_b. You can call this series combination R_{ab}. Figure 1-8 shows the network redrawn to show this series combination as a single resistance element.

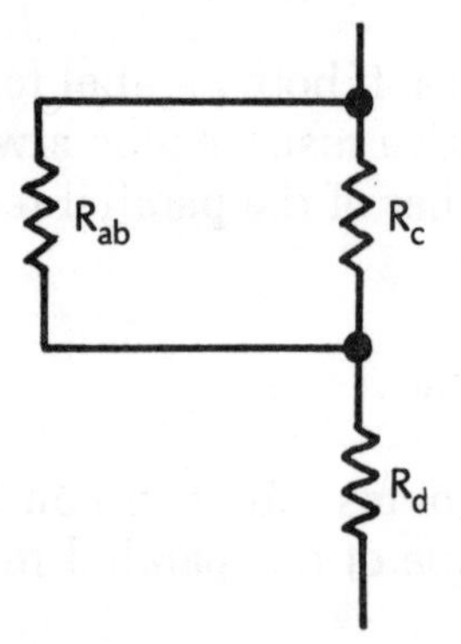

Fig. 1-8 This is the equivalent circuit for the circuit of Fig. 1-7 after solving for the series combination of R_a and R_b.

The next step is to find the parallel value of R_c and R_{ab}. Fig. 1-9 shows the resistance network redrawn once more to show this combined resistance (R_{abc}) as a single component. Now all you

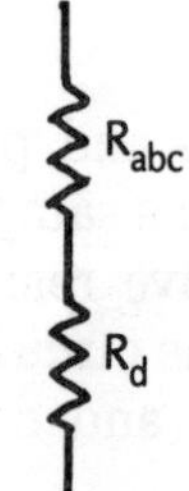

Fig. 1-9 This is the equivalent circuit for the circuit of Figs. 1-7 and 1-9 after solving for the parallel combination of R_{ab} and R_c.

have to do is find the series combination of the values for R_d and R_{abc}. This will be the total effective resistance (R_t) for the entire series/parallel resistance network as a whole.

I generally assume all small capacitors (under 1μF) to be inexpensive ceramic discs. There is nothing to stop you from using a more expensive, higher-grade component such as a mylar or a polystyrene capacitor. In most of these projects, there is no particular advantage in using higher-grade components. Once again, use whatever you have handy. In most cases, you can substitute values other than those specified in the parts lists. Generally, use a component with a value as close as possible to the specified value, except in cases where I encourage experimentation.

Capacitors generally are available in just a few standardized values. Oddball values might be necessary from time to time. You can combine multiple capacitances to create a desired value, just as with resistors. Like resistors, you can combine capacitors either in series or in parallel, or a combination of the two. The larger capacitors (1μF and up) used in these projects are standard electrolytic units.

The formulas for determining series and parallel capacitances are similar to those used for resistances, except they are reversed. That is, for capacitors in series, the formula is:

$$1/C_t = 1/C1 + 1/C2 + 1/C3 \ldots + 1/C_n$$

For capacitances in parallel, the values simply add:

$$C_t = C1 + C2 + C3 + \ldots C_n$$

I discuss any special notes on substituting parts in the text for the appropriate project(s).

SOCKETS

People working in electronics strongly disagree about the use of IC sockets. Some people swear by them, and others swear at them.

Some technicians insist that sockets should be used only for chips that are changed frequently by the user (such as ROMs containing different programming for various functions) or possibly to protect very expensive ICs. When a socket is used, the socket is soldered into the circuit, then the IC is inserted into the socket. There is no risk of damaging the chip by using too much heat when soldering. Others in the electronics field (myself included) recommend the regular use of all IC sockets, except in special, rare cases.

It might seem pretty silly to protect a 25-cent IC with a 50-cent socket, but what you are really protecting is your own time and sanity. If you make a mistake, apply a little too much heat, or if an IC has to be replaced for servicing at a later date, an IC socket can simplify the job immensely. Without a socket, you must desolder and then resolder each individual pin on the IC package, while carefully watching out for solder bridges and possible overheating. Frankly, I don't think that's worth the trouble, especially if such problems can be avoided so easily. Sockets don't add that much to the cost of a project, and they can head off a lot of grief and frustration if problems do arise. Think of IC sockets as a sort of insurance policy.

Those who oppose sockets often point out how easy it is to insert an IC into a socket backwards. When power is applied to the circuit, a backwards IC will be damaged almost certainly because the power-supply voltages will be fed to the wrong pins. However, it is just as easy to install an IC backwards even if a socket is not used. Either you work carefully and make it a practice of double-checking before applying power, or you don't. It's not the fault of the socket.

There are some special cases in which the use of IC sockets is undesirable. In equipment intended for field use, sockets might not be advisable. Such equipment is likely to be bounced around a lot. Direct soldering might be a good idea in such equipment to prevent a chip's pin from being bounced out of place.

A few (very few) high-frequency circuits can be disturbed by the slightly poorer electrical connections resulting from the use of a socket. These problematic circuits, however, are few and far between. In 99 percent of the circuits you're likely to work with (including all of those presented in this volume), using IC sockets will not create problems and could save you a lot of needless hassle.

HEAT SINKS

Although semiconductor components (such as transistors or integrated circuits) can handle moderate to large amounts of power, a heat sink generally is necessary to dissipate heat generated within the semiconductor device itself. If unprotected, a transistor or IC could literally self-destruct if it tries to pass too much current with no way for the heat generated to be carried away from the delicate semiconductor crystal.

A heat sink is nothing more than a thermal conductor designed to carry heat away from a temperature-sensitive device and dissipate it, usually into the surrounding air.

Most heat sinks are metallic shields (usually with fins to maximize the area exposed to the surrounding air) that are fitted over the component to be protected. Metal heat sinks of various shapes and sizes are available from many sources of electronics components. Figure 1-10 illustrates some typical heat sinks. When in doubt, use the next larger size. It is better to have too much heat sinking than too little.

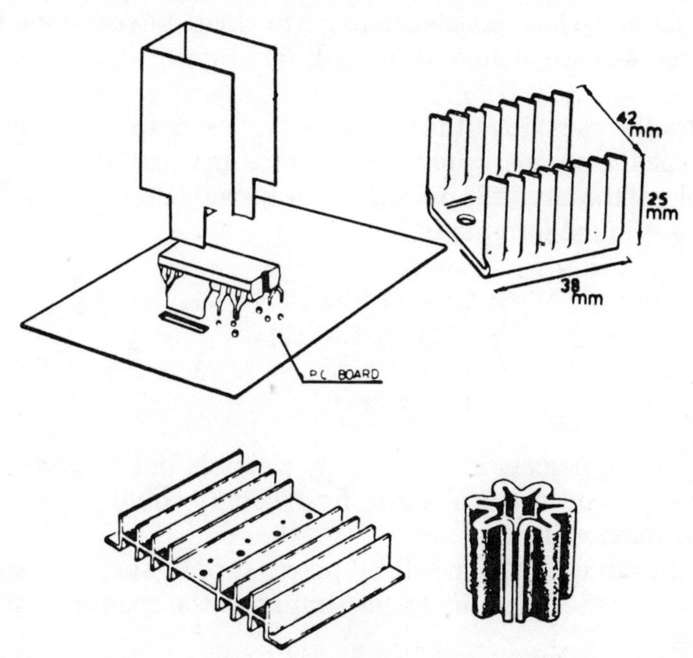

Fig. 1-10 Heat sinks are used to protect semiconductor components.

You can maximize heat transfer between the component and the actual heat sink by using a special heat-sink compound between the two. Heat-sink compound is available from Radio Shack stores and many other electronics suppliers.

On some printed circuit boards, you can include large copper pads to serve as simple heat sinks, if large amounts of power are not involved. ICs that are suitable for heat sinks of this type (including a number of amplifier ICs) often are fitted (by the manufacturer of the chip) with one or more tabs for soldering directly to the copper pads forming the heat sink.

You can calculate the area of copper (size of the pad) needed for a heat sink quite simply if you know the relevant circuit parameters. First, you must determine the maximum power that the heat sink must dissipate. Use the following formula:

$$\text{power (in watts)} = 0.4 \times (V_S^2/8R_L) + (V_S \times I_d)$$

where V_S is the maximum supply voltage

I_d is the quiescent drain current (in amperes) under the most adverse conditions (worst-case figure)

R1 is the load resistance (e.g., the loudspeaker impedance in the case of an audio-amplifier circuit).

Strictly speaking, the value of V_S used in this equation should be the power-supply voltage, plus an extra 10 percent. For example, if the circuit is powered by a 12-volt battery, the value of V_S for use in the formula is:

$$\begin{aligned} V_S &= 12 + (10\% \text{ of } 12) \\ &= 12 + (0.1 \times 12) \\ &= 12 + 1.2 \\ &= 13.2 \text{ volts} \end{aligned}$$

This extra 10 percent allows for any possible fluctuations in the power level. For instance, a very fresh battery could put out more than its rated voltage.

If the circuit has a stabilized power supply, such as a voltage regulator, then V_S can be taken simply as the regulated supply voltage.

You can usually find the quiescent drain current (I_d) among the parameters in the manufacturer's spec sheet for the chip. This value depends on the supply voltage used in the circuit. On most

IC spec sheets, I_d figures often are quoted for "typical" and "maximum." In this case, use the maximum value in the equation.

Figure 1-11 illustrates the relationship between the power to be dissipated and the required copper area based on a maximum ambient temperature of 55° C. This is a safe upper limit for most common semiconductor devices.

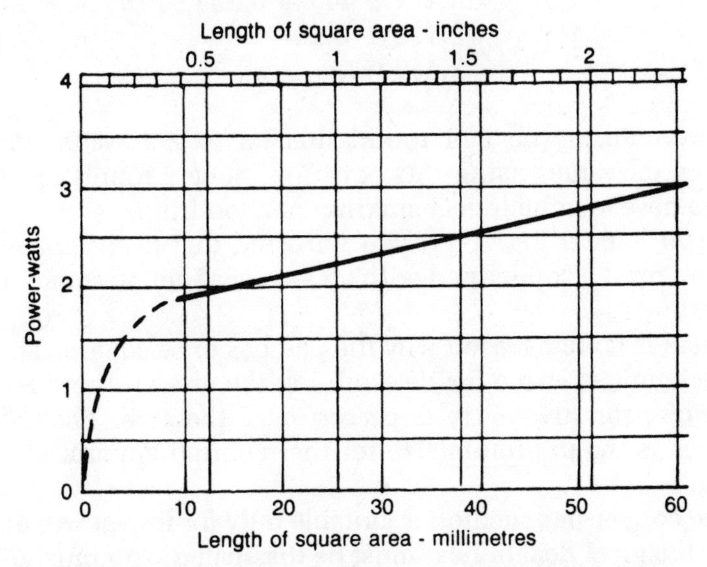

Fig. 1-11 *This graph illustrates the relationship between the power dissipated and the required area of a copper-trace heat sink.*

As an example of how this graph could be used, let's assume that the supply voltage in a circuit is 12 volts, and the load resistance is 4 ohms. To keep things simple here, I will assume we only need to protect a single IC with heat sinking. The maximum quiescent current drain quoted by the manufacturer for this particular device is 20 mA (0.2 amp) at 12 volts. I will further assume that the supply voltage is not regulated in this particular circuit.

Because the supply voltage is unregulated, you need to include our 10 percent "fudge factor" in our value for V_S:

$$V_S = 12 + 1.2$$
$$= 13.2 \text{ volts}$$

You now have all the information you need to determine the power which needs to be dissipated. You just plug your values into the equation:

$$\begin{aligned}\text{Power (in watts)} &= 0.4 \times (V_S^2/8R_L) + (V_S \times I_d)\\ &= 0.4 \times (13.2^2/(8 \times 4)) + (13.2 \times 0.02)\\ &= 0.4 \times (174.24/32) + 0.264\\ &= 0.4 \times 5.445 + 0.264\\ &= 2.178 + 0.264\\ &= 2.442 \text{ watts}\end{aligned}$$

For convenience, you can round this off to 2.5 watts. When rounding off values using this equation, always round up. It is better to have too much heat sinking than too little.

If you look at Fig. 1-11, you will find that to dissipate 2.5 watts you need a copper pad with an area of about 40 square millimeters.

There is no real reason why the pad has to be square. This is just the simplest shape for determining the area. If there is any margin of error, always try to overestimate the area. The calculated area is the minimum size for the required amount of heat sinking.

The copper-pad method is suitable only for low-power dissipation. If a great deal of heat must be dissipated, you must use a metal heat sink, like the ones shown back in Fig. 1-10.

CUSTOMIZING THE PROJECTS

As you read about the projects in this book, you are likely to feel that some come close to your individual needs, but aren't quite right. You should feel free to customize any or all of the projects to suit your own individual needs.

Often you can ignore the original intended application (as described in the text) and try to determine if the circuit can be put to work for the application you have in mind. Use your imagination. Frequently a small change in a circuit will produce a totally different device. For example, a circuit that responds to changes in lighting levels could be made to respond to changes in temperature, perhaps by just substituting a thermistor for the original photoresistor. In many cases, you can achieve considerable customization simply by changing a sensor or the input signal. Sometimes you can combine two or more simple projects to create a more complex system.

Block diagrams can be a big help when designing customized systems. Determine what each stage in the circuit needs to do, then find a circuit that serves that function. This is far easier than designing a complex circuit from scratch.

Actually, you can break down most (if not all) complex circuits built by either commercial manufacturers or hobbyists into several relatively simple stages, or subcircuits. As always, use your imagination, and take your projects as far as you can.

I very strongly recommend that you breadboard any circuit changes before permanently soldering them. Occasionally, as many hobbyists and technicians have learned to their grief, what works on paper might not work in quite the same way in an actual circuit. It's better to find any problems or surprises early on, when it is easy to make additional changes and reuse the components. Waiting until you've got the whole circuit soldered together before you test out your modifications is an open invitation to frustration.

TWELVE-VOLT POWER SOURCES

All of the projects in this book require a 12-volt dc power supply. You can supply the necessary voltage in several ways to the circuits.

Batteries

The most obvious and simplest source of dc power is a battery pack. Most standard batteries (AAA, AA, C, and D) put out 1.5 volts. To make up a 12-volt battery, you need to use eight 1.5-volt cells in series. To save space, you can use a 9-volt transistor radio battery in series with four 1.5-volt cells (probably AA, or AAA).

A few large lantern batteries supply 12 volts all by themselves. However, 6-volt lantern batteries are more common. To make up 12 volts dc, just wire a pair of 6-volt lantern batteries in series.

Automobile Power

Most automobiles use 12-volt batteries. This voltage can be tapped off at various points and used to power electronic circuits.

The simplest approach is to take the 12 volts dc from the cigarette lighter. Special adapters for this purpose are available from Radio Shack and many other dealers.

Chapter 2 features several projects specifically designed for automotive use. I intended all of them to be wired into the car's electrical system.

AC-to-DC Converters

If you will be using a project exclusively in an area where ac house current is available, it makes sense to tap this power source. It is certainly cheaper than batteries.

However, these projects are designed to run off of 12 volts dc, and house current is 110 volts ac. Obviously, you need a conversion circuit.

A simple "quick and dirty" 12-volt power-supply circuit appears in Fig. 1-12. Table 1-1 shows a parts list for this circuit. You can use this power supply to run a noncritical circuit.

Some circuits, especially those involving digital circuitry, require a regulated dc supply voltage. Figure 1-13 shows a regulated 12-volt power-supply circuit, and the parts list appears in Table 1-2.

Always use a fuse or circuit breaker with any ac-powered circuit. These safety devices interrupt the ac input to the circuit if the current drawn becomes too great. If you don't use a fuse or

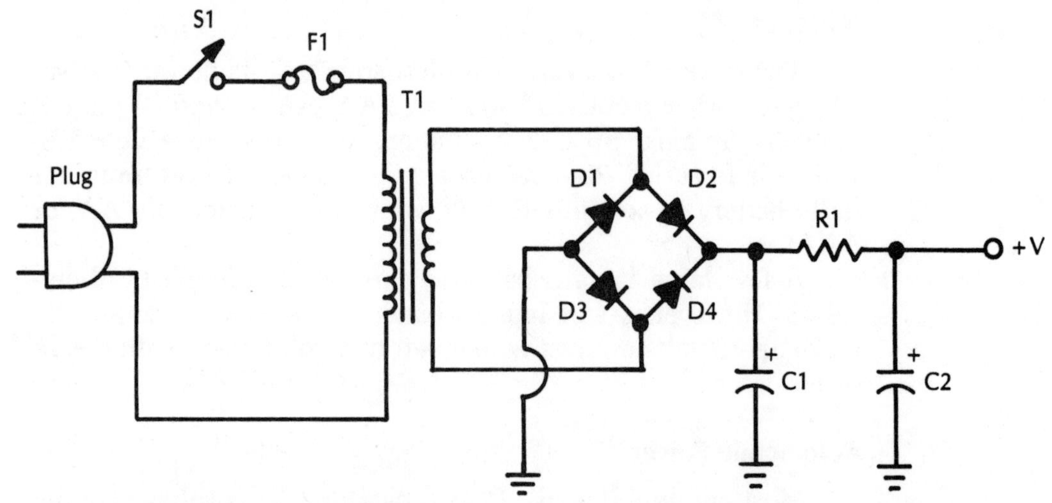

Fig. 1-12 This "quick and dirty" power-supply circuit can be used to run the projects in this book.

Table 1-1 Parts List for the Power-Supply Circuit of Fig. 1-12.

Part	Component
D1 – D4	Diode (1N4003, or similar)
C1, C2	470μF 30-volt electrolytic capacitor
R1	1K, 1/2-watt (or better) resistor
T1	Power transformer secondary rated for at least 15 – 18V ac, 150% of expected current load
S1	SPST switch
F1	Fuse (rated to suit intended load)

circuit breaker, the result could be a painful and dangerous, and possibly even fatal, electrical shock.

You should select the fuse or circuit-breaker rating for the individual circuit to be powered. Estimate the maximum amount of current the device should normally draw, and add about 10 percent as a "fudge factor." For example, if a circuit normally draws up to 1.1 amp, adding 10 percent gives us 1.21 amps. In this case, use a 1.5-amp fuse, which is the nearest standard value.

Do not use too large a fuse (or circuit breaker). Never replace the existing fuse with a higher-rated unit. If the fuse blows repeatedly, or if the circuit breaker trips repeatedly, something is wrong. Disconnect power and troubleshoot the problem. Do <u>not</u> bypass the fuse or circuit breaker. Do <u>not</u> use a higher-rated device. If you use too large a current rating for the fuse or circuit breaker, some component in the circuit (usually a relatively expensive transistor or IC) will blow to protect the fuse or circuit breaker. Obviously, there isn't much point to that. Don't defeat the purpose of the safety device.

Make sure that all conductors carrying ac house current are well insulated. It should not be possible for anyone to touch accidentally a live conductor.

Follow any special safety precautions described for individual projects.

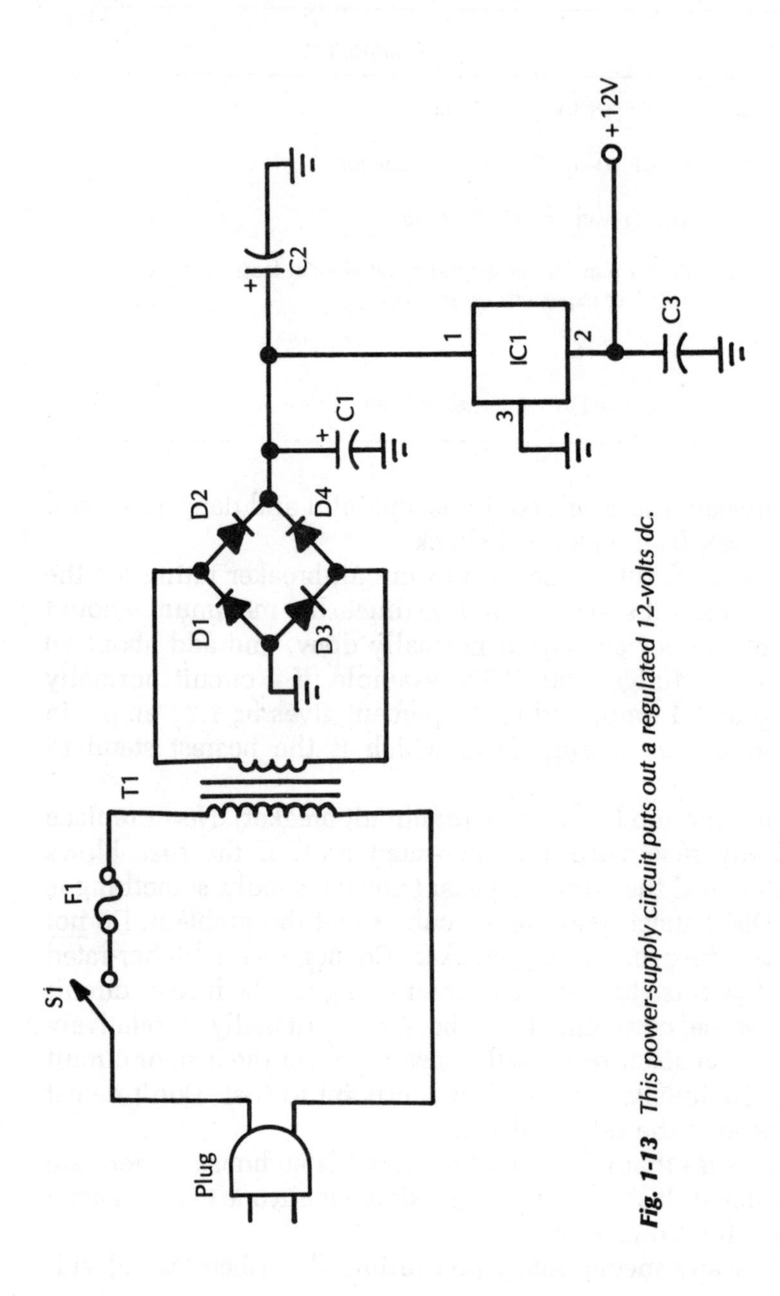

Fig. 1-13 *This power-supply circuit puts out a regulated 12-volts dc.*

Table 1-2 Parts List for the Power-Supply Circuit of Fig. 1-13. Regulated

Part	Component
IC1	7812 +12-volt voltage regulator
D1 – D4	Diode (1N4003, or similar)
C1, C2	470μF, 30-volt electrolytic capacitor
C3	0.5μF capacitor
T1	Power transformer, secondary rated for at least: 15 – 18V ac, 150% of expected current load
S1	SPST switch
F1	Fuse (rated to suit load)

2 ❖ Automotive Projects

THE BATTERY IN YOUR CAR IS AN EXCELLENT AND CONVENIENT source of 12-volts dc. This chapter features several projects intended for automotive use. These circuits are intended to add to your safety, enjoyment, and convenience when in your car.

I assume a negative ground system in all of these projects. If you happen to have a vehicle with a positive-ground electrical system, you must reverse all of the polarities in these circuits.

PROJECT 1: EMERGENCY FLASHER

If you've ever had an accident, a flat tire, or some other problem at night or during stormy weather, you know visibility can be a real problem. Oncoming vehicles might not be able to see the danger. They could collide with a stalled vehicle.

It is a good idea to carry an emergency flasher in your car at all times. In case of a problem, set out the flasher and turn it on. It produces a bright, flashing light. The signal is impossible to miss.

Figure 2-1 shows an emergency-flasher circuit. The parts list for this project appears in Table 2-1.

You adjust potentiometer R1 to determine the flash rate. Using the component values called for in the parts list, you can adjust the flash rate from about 35 to over 150 flashes per minute.

You can operate the emergency-flasher circuit from the car's battery. It is a good idea to carry a 12-volt lantern battery along with the emergency flasher in case the emergency involves a dead auto battery.

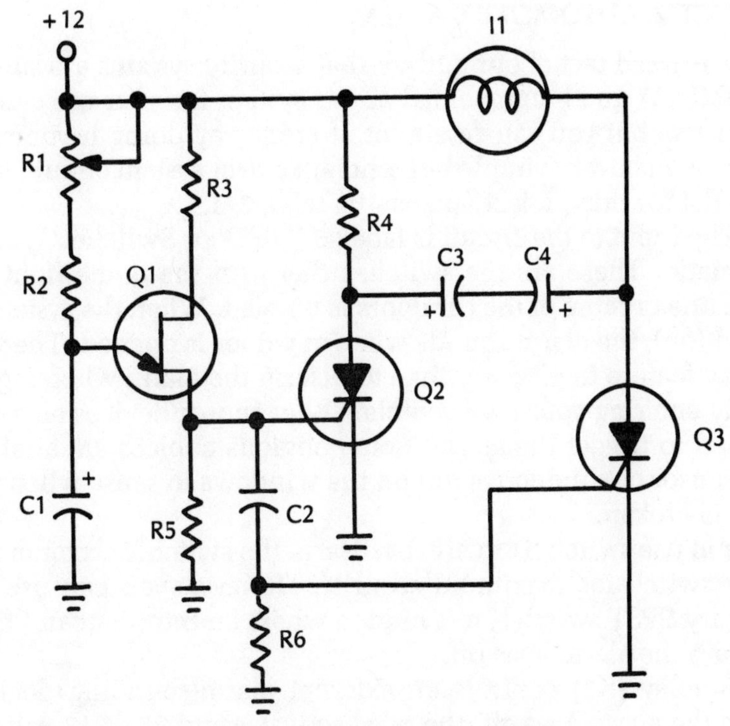

Fig. 2-1 Project 1: Emergency Flasher.

Table 2-1 Parts List for the Emergency Flasher of Project 1.

Part	Component
Q1	UJT (2N2646, or similar)
Q2, Q3	SCR (C1064, or similar)
I1	12-volt lamp (GE #1073, or similar)
C1	4.7-μF, 25-volt electrolytic capacitor
C2	0.01-μF capacitor
C3	100-μF, 25-volt electrolytic capacitor
R1	250k potentiometer
R2	3.9k, 1/4-watt resistor
R3	220-ohm, 1/4-watt resistor
R4	2.2k, 1/4-watt resistor
R5	10-ohm, 1/2-watt resistor
R6	1k, 1/2-watt resistor

PROJECT 2: AUTOMOBILE ALARM

It is a sad fact of our culture that security systems are such a necessity. A good commercial alarm system for your car can be expensive, but you can save a lot of money by doing it yourself. Figure 2-2 shows a simple but efficient alarm-system circuit. The parts list for this project appears in Table 2-2.

The input to the circuit is labeled "To Door Switches" in the schematic. These are the switches that turn the dome light on when one or more of the car doors is opened. When the system is armed (on), the alarm sounds when any door is opened. There is a delay feature to give you time to disarm the alarm when legitimately entering your own vehicle. Other sensor devices also can be used to trigger the alarm. Some obvious choices are a vibration sensor or conductive foil on the windows to sense when the glass is broken.

You use switch S1 to arm or disarm the switch. I recommend a key switch for maximum security. However, you can use an ordinary SPST switch if it is hidden where an intruder can't find it before the alarm goes off.

A relay (K1) controls an external alarm-sounding device. When the alarm goes off, the relay activates and feeds 12 volts to the external alarm-sounding device. This device will probably be the vehicle's horn or a siren in most installations.

The main control functions of the circuit are performed by a pair of timers. IC1 contains both of these timers.

The timing components for timer A are resistor R4 and capacitor C2. This timer controls the delay time. The delay is the time between the time when a door is opened (or other sensor device is triggered) and when the alarm actually goes off. This delay is necessary to provide time to open arming switch S1. With the component values given in the parts list, the delay time is approximately 15 seconds. To increase the delay time, simply increase the value of either R4 or C2, or both. Similarly, reducing the value of either or both of these components reduces the delay time.

Timer B controls how long the alarm continues to sound. The timing components here are resistor R11 and capacitor C5. With the values given in the parts list, the alarm sounds for about 90 seconds. The alarm then shuts itself off and automatically

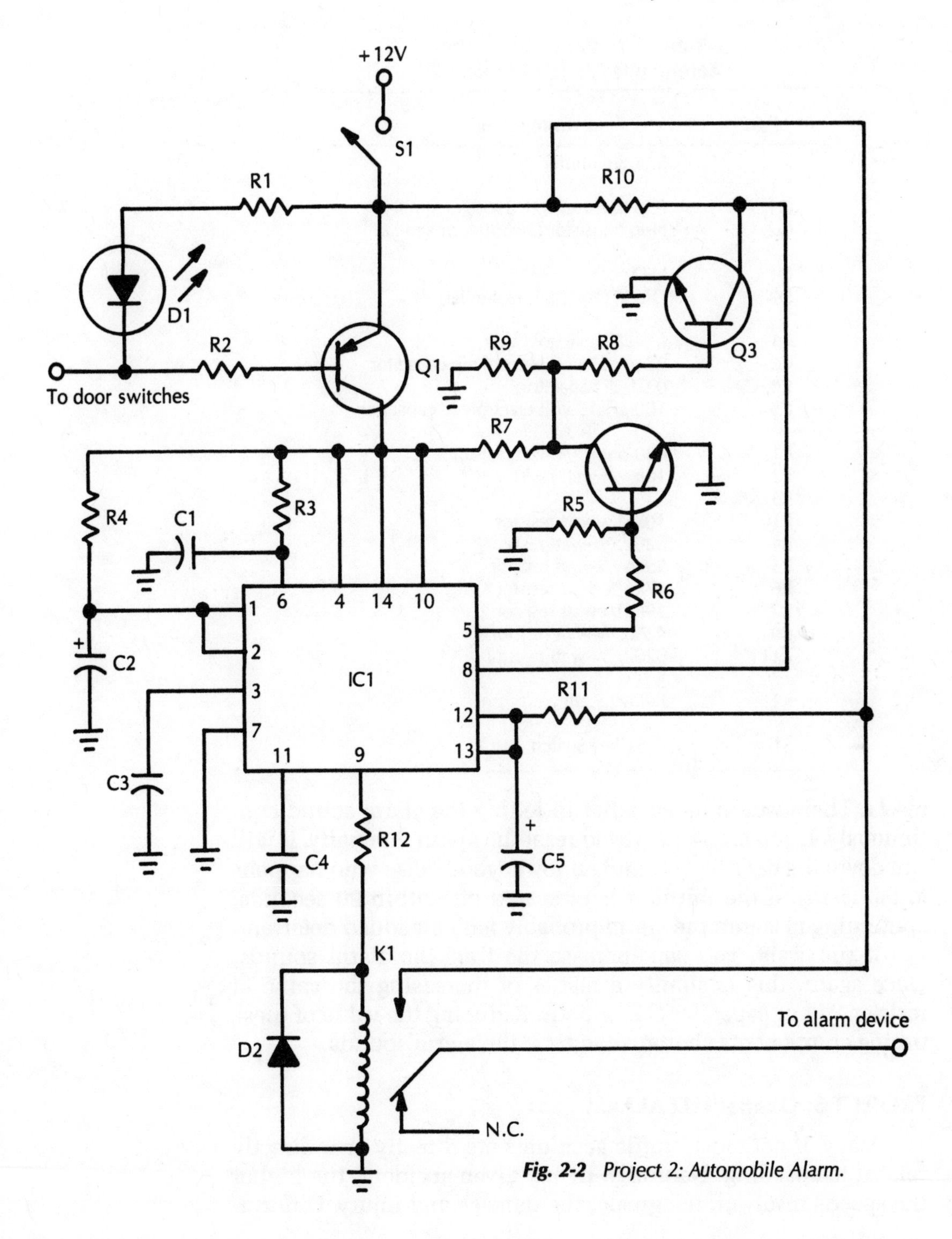

Fig. 2-2 Project 2: Automobile Alarm.

Table 2-2 Parts List for the Automobile Alarm of Project 2.

Part	Component
IC1	556 dual timer
Q1	Pnp transistor (2N3906, or similar)
Q2, Q3	Npn transistor (2N3904, or similar)
D1	LED
D2	Diode (1N4004, or similar)
C1	0.1-μF capacitor
C2	22-μF, 15-volt electrolytic capacitor
C3, C4	0.01-μF capacitor
C5	100-μF, 15-volt electrolytic capacitor
R1	680-ohm, 1/4-watt resistor
R2	18k, 1/4-watt resistor
R3, R8, R10, R12	10k, 1/4-watt resistor
R4	620k, 1/4-watt resistor
R5	5.6k, 1/4-watt resistor
R6	47k, 1/4-watt resistor
R7	39k, 1/4-watt resistor
R9	4.7k, 1/4-watt resistor
R11	820k, 1/4-watt resistor
K1	Relay to suit alarm
S1	SPST key switch

resets. There would be no point in letting the alarm sound continuously. If you aren't nearby to reset the alarm manually, it will run down the car's battery and annoy anybody else who happens to be nearby. If the intruder isn't scared off within 90 seconds, continuing to sound the alarm probably isn't an added deterrent.

If you want, you can increase the time the alarm sounds. Once again, this is simply a matter of increasing the value of resistor R11 or capacitor C5, or both. Reducing the value of these timing components shortens the time the alarm sounds.

PROJECT 3: OVERSPEED ALARM

Many, if not most, traffic accidents are directly or indirectly related to speeding. Naturally, in any given accident, the higher the speeds involved, the greater the damage and injury. Unfortunately, it is very easy to start driving too fast. When you are concentrating on where you're going and the other cars around you, it is difficult to keep an eye on the speedometer.

Sometimes ads for radar detectors claim that these devices promote driving safety by making the driver more aware of his speed, but this claim is clearly false. A radar detector just warns the driver when he is being officially watched. The driver should stay within the legal speed limit at all times, whether he is being watched or not. Radar detectors actually encourage speeding because they reduce the chances of receiving a ticket for breaking the law.

If you really are interested in safety, an overspeed alarm will do far more good than a radar detector. The circuit shown in Fig. 2-3 sounds a tone and alerts you whenever you exceed a preset speed. The parts list for this project appears in Table 2-3.

The input to this circuit is in the form of pulses from the distributor points. Transistor Q1 and its associated components convert the input pulses into a dc voltage that is proportional to the engine's rpm.

IC1A and IC1B basically act as a comparator/Schmitt trigger. When the dc voltage exceeds a specific preset value (set by potentiometer R4), these gates enable the gated oscillator made up of IC1C, IC1D, and their associated components. The signal from this oscillator is fed to the speaker, producing an attention-getting tone. This tone continues to sound as long as you exceed the preset speed. Reducing the speed by a few miles per hour causes the alarm to shut down and reset itself.

With this project installed in your car, you'll never accidentally speed again. You can still speed if you really want to, of course, but this device lets you know about it, even when you aren't paying attention to the speedometer.

An overspeed alarm like this is best if it is set for the maximum legal highway speed in your area. Highway driving is where accidental speeding is most likely (and most dangerous).

PROJECT 4: WINDSHIELD WIPER DELAY

Most standard windshield wiper mechanisms have just a single speed, or at most, two speeds. In a light drizzle, a windshield wiper operating at full speed can be annoying and could leave vision-blurring streaks across the glass.

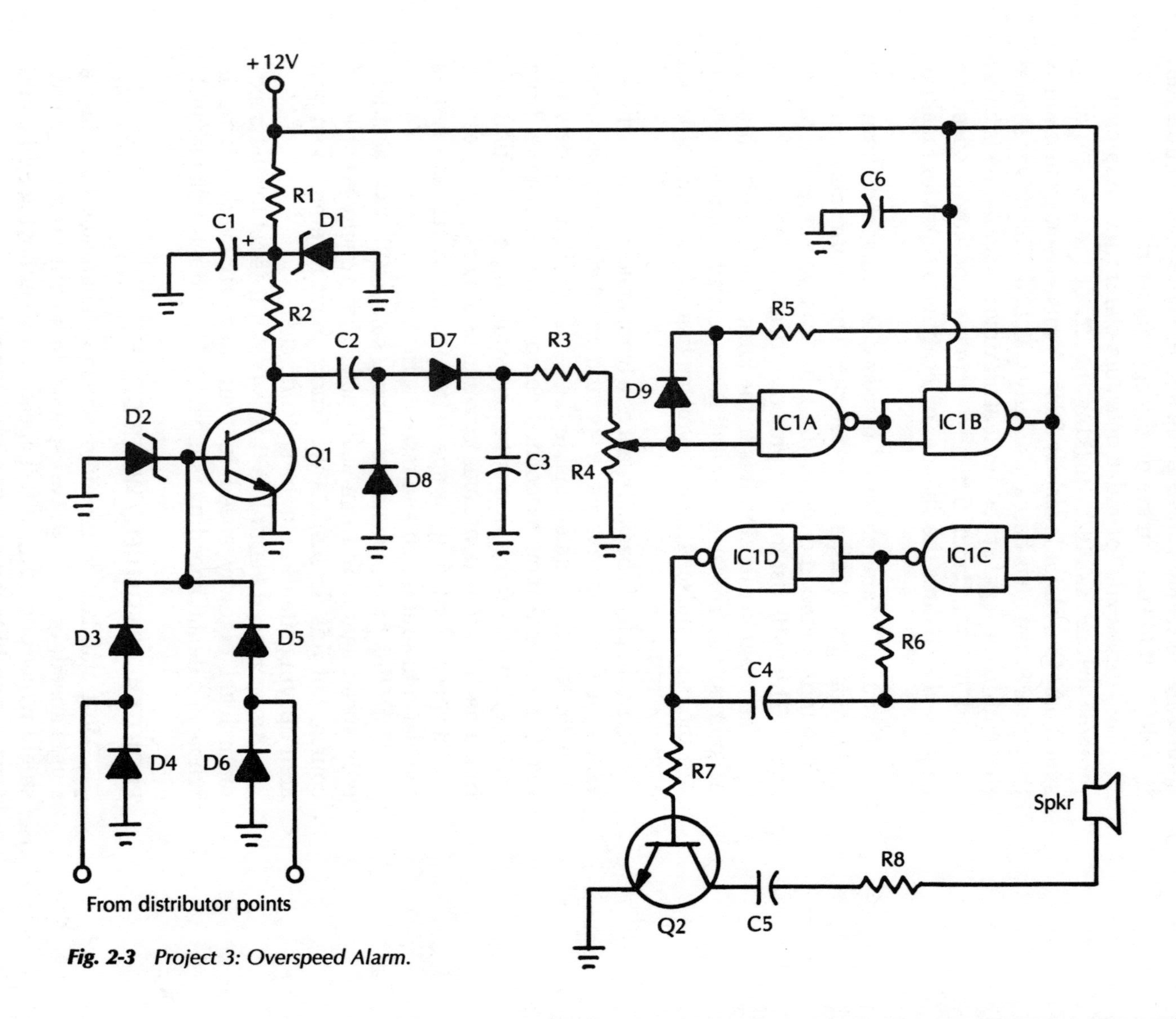

Fig. 2-3 *Project 3: Overspeed Alarm.*

Table 2-3 Parts List for the Overspeed Alarm of Project 3.

Part	Component
IC1	CD4011 quad NAND gate
Q1	Pnp transistor (2N3904, or similar)
Q2	Pnp transistor (TIP3055, or similar)
D1	Zener diode—9 volts
D2	Zener diode—4.7 volts
D3–D9	Diode (1N4148, or similar)
C1	100-μF, 15-volt electrolytic capacitor
C2, C3, C6	0.1-μF capacitor
C4, C5	0.01-μF capacitor
R1	1k, 1/4-watt resistor
R2	4.7k, 1/4-watt resistor
R3	47k, 1/4-watt resistor
R4	250k potentiometer
R5	10-megohm, 1/4-watt resistor
R6, R7	22k, 1/4-watt resistor
R8	270k, 1/4-watt resistor
Spkr	Small speaker

The circuit shown in Fig. 2-4 allows you to slow down the windshield wipers to a more convenient speed. Table 2-4 shows the parts list for this project.

IC1, a 555 timer, is the heart of this circuit. This chip is connected here as a pulse generator, or *astable multivibrator.* The output pulses open and close the relay (K1) contacts. The wipers can operate only when the relay contacts are closed. By rapidly opening and closing these contacts, the power reaching the wipers is reduced, effectively decreasing their operating speed.

There are two speed controls in this circuit. R1 is a trimpot. You calibrate this control to set the minimum delay. The main manual speed control is potentiometer R3. Depending on the settings of R1 and R3, you can set the wiper speed anywhere from about one wipe per second to about one wipe every 20 seconds.

Switch S1 is an optional bypass switch. When you close this switch, the delay circuit is bypassed, and the windshield wipers operate in their normal (nondelayed) manner.

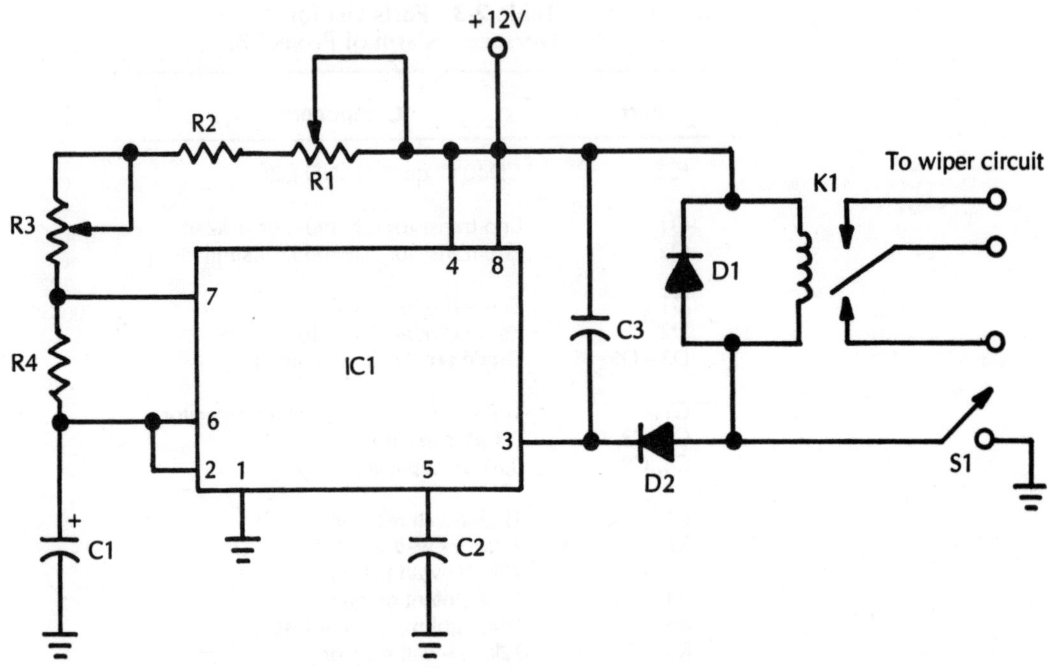

Fig. 2-4 Project 4: Windshield Wiper Delay.

Table 2-4 Parts List for the Windshield Wiper Delay of Project 4.

Part	Component
IC1	555 timer
D1, D2	Diode (1N4004, or similar)
C1	100-μF, 15-volt electrolytic capacitor
C2	0.01-μF capacitor
C3	0.1-μF capacitor
R1	25k trimpot
R2	1.2k, 1/4-watt resistor
R3	250k potentiometer
R4	3.9k, 1/4-watt resistor
K1	Relay, 150-ohm coil
S1	SPST switch

PROJECT 5: TACHOMETER

If you want to tune up your car's engine yourself, you need a tachometer. Figure 2-5 shows a circuit for a simple tachometer. The parts list for this project appears in Table 2-5.

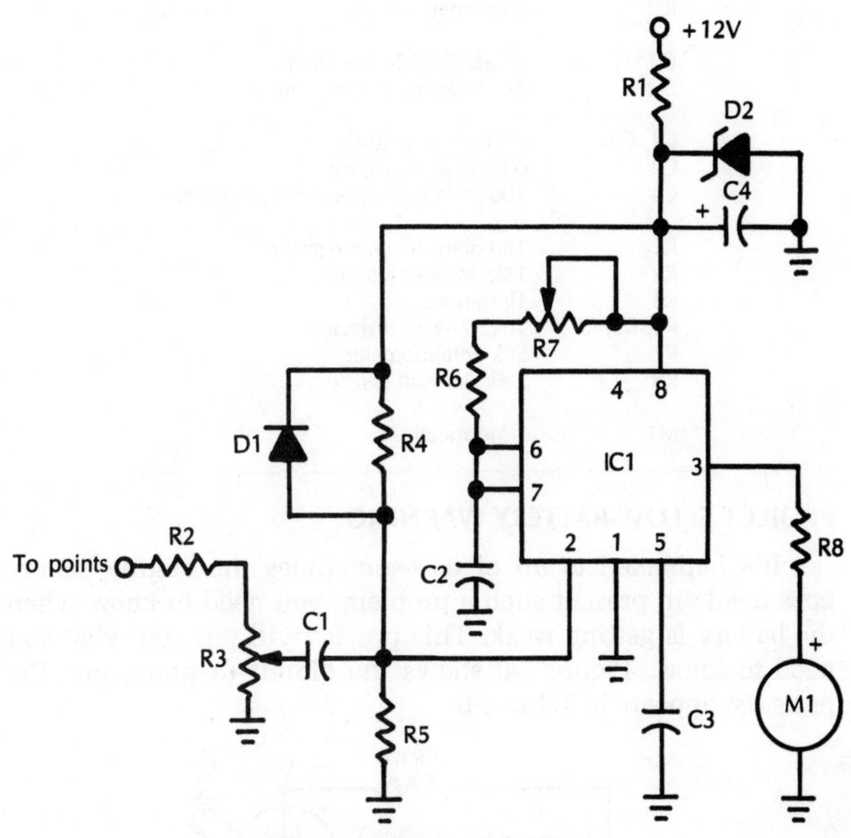

Fig. 2-5 *Project 5: Tachometer.*

The input is taken from the distributor. In this project, you make the connection to the low-tension (12-volt) side of the distributor.

The distributor signal is in the form of pulses. These pulses are converted within this circuit into a dc voltage that is proportional to the engine's rpm.

Trimpot R3 is a calibration control. If you can borrow another tachometer from someone else, you can calibrate this unit for the same reading under identical conditions.

Table 2-5 Parts List for the Tachometer of Project 5.

Part	Component
IC1	555 timer
D1	Diode (1N4003 or similar)
D2	Zener diode, 8.2-volt rating
C1, C3	0.01-μF capacitor
C2	0.0056-μF capacitor
C4	100-μF, 15-volt electrolytic capacitor
R1	180-ohm, 1/4-watt resistor
R2	15k, 1/4-watt resistor
R3	1k trimpot
R4, R5, R6	10k, 1/4-watt resistor
R7	50k potentiometer
R8	5.6k, 1/4-watt resistor
M1	1-mA meter

PROJECT 6: LOW-BATTERY WARNING

It's happened to all of us—sometimes the engine battery goes dead. To prevent such a problem, you need to know when the battery is getting weak. This project will tell you what you need to know. Figure 2-6 shows the circuit diagram, and the parts list appears in Table 2-6.

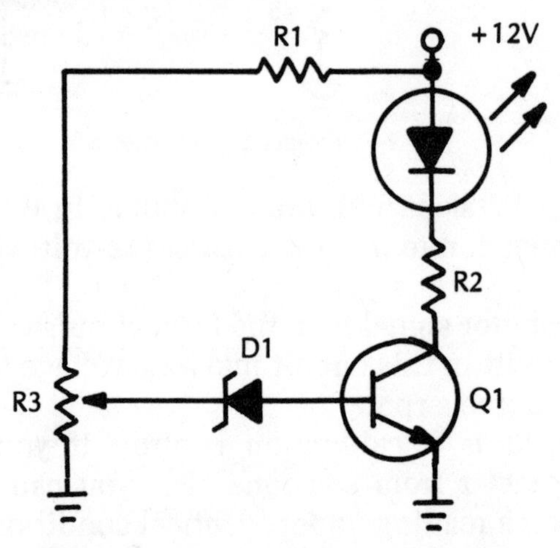

Fig. 2-6 Project 6: Low-Battery Warning.

Table 2-6 Parts List for the Low-Battery Warning of Project 6.

Part	Component
Q1	Npn transistor (2N3904, or similar)
D1	Zener diode, 6.2 volt rating
D2	LED
R1	1.2k, 1/4-watt resistor
R2	680-ohm, 1/4-watt resistor
R3	10k trimpot

This project is really quite simple. When the battery voltage drops below a preset value, the LED lights up. The trigger point can be set by potentiometer R3. For most applications, a trimpot is used for R3. Once you set the trigger voltage, you probably won't want to change it.

Of course, this project is not limited to automotive use. You can employ it to monitor any 12-volt dc power source.

PROJECT 7: HEADLIGHT DELAY TIMER

Have you ever parked your car in a dark garage and then had to fumble around to find the light switch? You could leave the headlights on until you turn on the garage light, and then reach back in the car to turn the headlights off, but this is an inelegant solution at best. Besides, it's all too easy to forget and leave the headlights on all night, which will give you a good chance of finding a dead battery in the morning.

The circuit shown in Fig. 2-7 is a handy, automated headlight control that will prevent such a problem. The parts list for this project appears in Table 2-7.

When activated, this circuit turns the headlights on for a predetermined period of time. After the circuit times out, it automatically turns the headlights back off. You don't have to worry about it.

Capacitor C1 and potentiometer R1 determine the delay time (how long the headlights stay lit). You adjust R1 for the desired time period. In most applications, it makes sense to use a trimpot instead of a manual front-panel control for R1. Depending on the

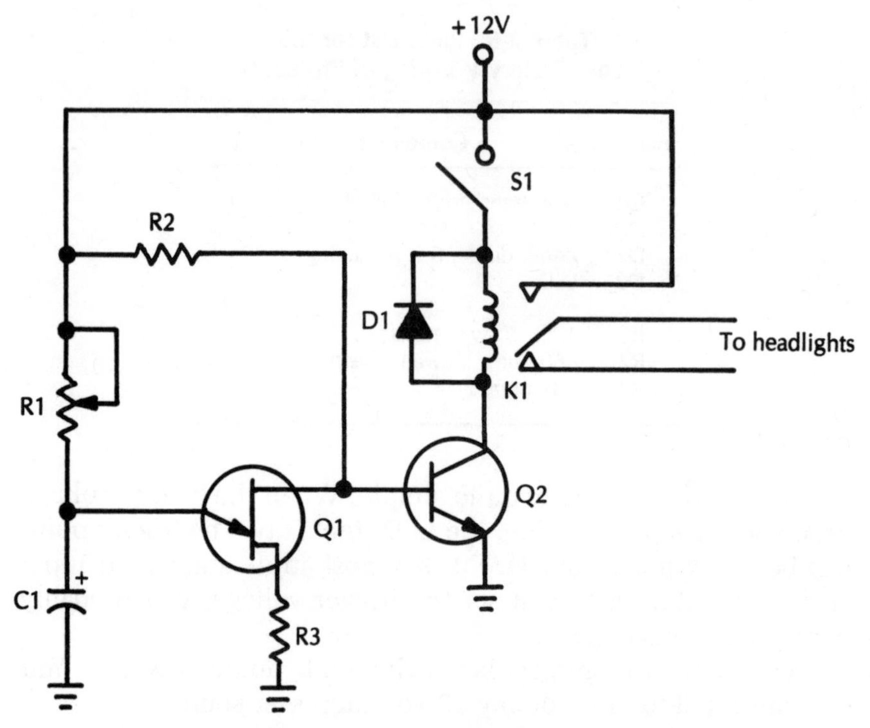

Fig. 2-7 Project 7: Headlight Delay Timer.

Table 2-7 Parts List for the Headlight Delay Timer of Project 7.

Part	Component
Q1	UJT (2N2646, or similar)
Q2	Npn transistor (HEP50, GE-20, or similar)
D1	Diode (1N4003 or similar)
C1	10-μF, 15-volt electrolytic capacitor
R1	10k potentiometer
R2	330-ohm, 1/4-watt resistor
R3	1k, 1/4-watt resistor
K1	Relay (coil—180 ohms)
S1	SPST switch

setting of R1, the lights will remain on for a period of a minute or two.

Switch S1 is a normally open pushbutton type (SPST). Briefly closing this switch activates the timer. This project does not interfere with the ordinary operation of the headlights.

3 ❖ Timing Projects

A COMMON CLASS OF APPLICATIONS FOR ELECTRONICS INVOLVES timing. Certain events (triggered by electrical signals) might need to occur at a certain time, or after a specific delay after some other event. Several events might need to occur in a specific sequence with predetermined intervals between them. This chapter features several sequencer and time-delay projects for a wide variety of applications.

The projects presented in this chapter also include timekeeping devices for measuring the passage of time, including a complete digital-clock project.

PROJECT 8: TWO-STAGE SEQUENTIAL TIMER

A 555 timer (monostable multivibrator) circuit is handy for triggering an event at a specific time after a control signal is received. In a more complex system, you might need to initiate a series of events from a single control pulse; for example, event B might need to be triggered after a longer delay than event A.

A circuit for accomplishing this appears in Fig. 3-1. Table 3-1 shows a suitable parts list for this project. You should experiment with other component values.

This circuit has two outputs, labelled A and B in the schematic diagram. When a trigger pulse is received, output A turns on for a period determined by the values of resistor R2 and capacitor C1. After this output times out and switches off, output B turns on for a period determined by the values of resistor R3 and capacitor C5. The formulas for calculating the outputs times are as follows:

$$\text{time A} = 1.1 \times R2C1$$
$$\text{time B} = 1.1 \times R3C5$$

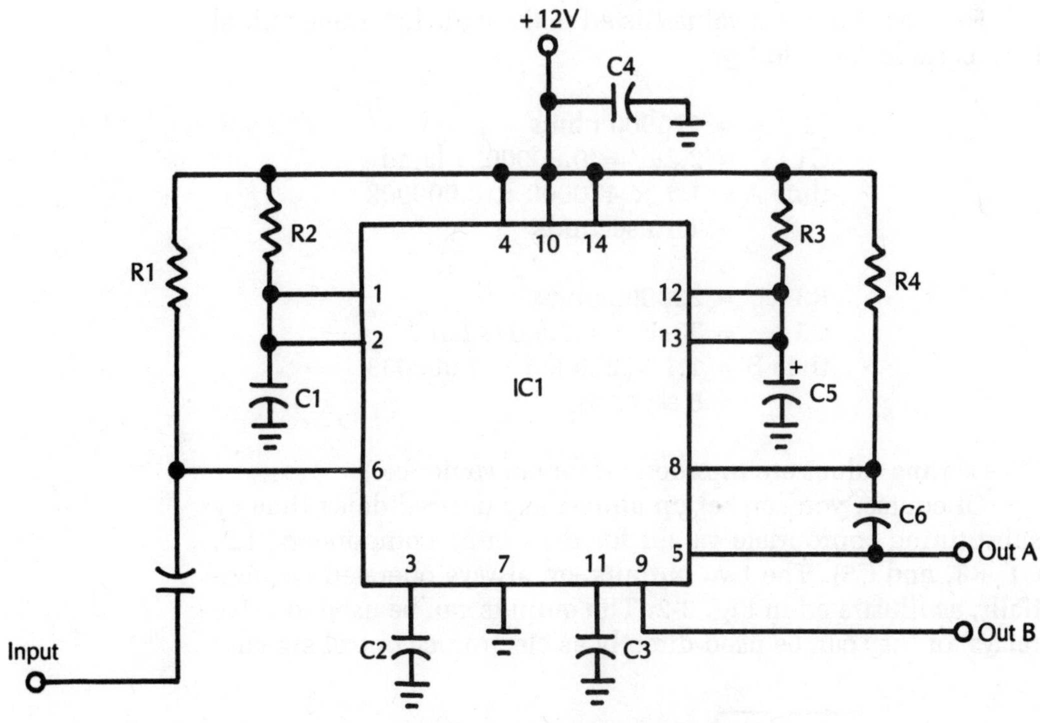

Fig. 3-1 *Project 8: Two-Stage Sequential Timer.*

Table 3-1 Parts List for the Two-Stage Sequential Timer of Project 8.

Part	Component
IC1	556 dual timer
C1	2.2-μF, 15-volt electrolytic capacitor*
C2, C3, C6	0.01-μF capacitor
C4	0.1-μF capacitor
C5	33-μF, 15-volt electrolytic capacitor*
R1, R4	10k, 1/4-watt resistor
R2	470k, 1/4-watt resistor*
R3	220k, 1/4-watt resistor*

*Timing component—see text

For the component values listed in the parts list, the nominal time periods are as follows:

$$\begin{aligned}
\text{R2} &= 470000 \text{ ohms} \\
\text{C1} &= 2.2\mu\text{F} = 0.0000022 \text{ farad} \\
\text{time A} &= 1.1 \times 470000 \times 0.0000022 \\
&= 1.15 \text{ seconds}
\end{aligned}$$

$$\begin{aligned}
\text{R3} &= 220000 \text{ ohms} \\
\text{C5} &= 33\mu\text{F} = 0.000033 \text{ farad} \\
\text{time B} &= 1.1 \times 220000 \times 0.000033 \\
&= 8 \text{ seconds}
\end{aligned}$$

(These time values are rounded off for convenience.)

Of course, you can set up almost any desired delay time by substituting appropriate values for the timing components (R2, C1, R3, and C5). The two outputs are always operated sequentially, as illustrated in Fig. 3-2. The outputs can be used to drive relays, or they can be used directly as electronic control signals.

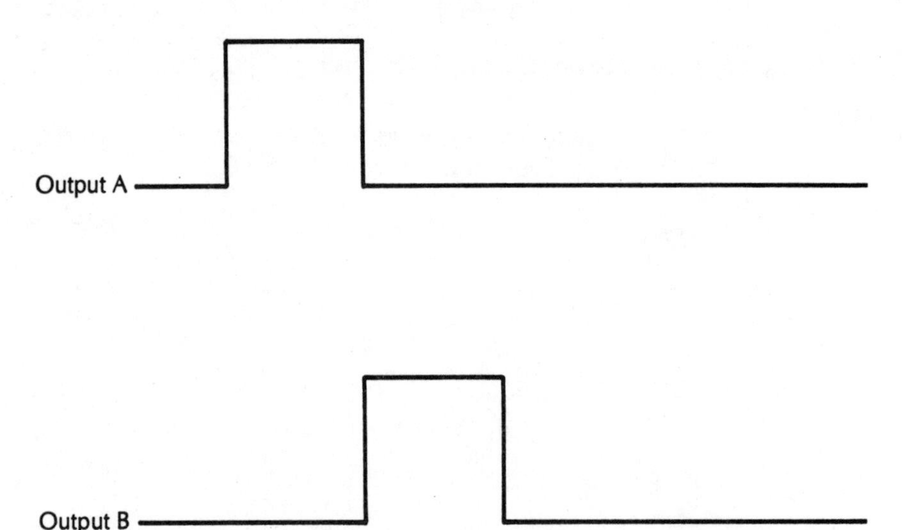

Fig. 3-2 *The two outputs of Project 8 are always operated sequentially.*

PROJECT 9: FOUR-STEP SEQUENCER

In many applications, certain events need to be triggered sequentially (one after another). Digital circuitry is well suited for this type of application. Most digital-sequencer circuits are based on some sort of counter being incremented by a clock (oscillator).

The circuit shown in Fig. 3-3 is a repeating four-step sequencer. Table 3-2 shows the parts list. Each of the four outputs activate sequentially in the following cyclic pattern:

1
2
3
4
1
2
3
4
1

and so forth.

The clock (oscillator) is built around IC1. This is a 7555 timer. The 7555 is a CMOS version of the popular 555 timer chip. You could substitute an ordinary 555 IC in place of the 7555. The two devices are pin-for-pin compatible. This substitution requires no changes in the wiring or circuitry.

You wire the timer in the astable mode to serve as the system clock. You can adjust the clock frequency by potentiometer R2 or by changing the value of capacitor C1. The larger this capacitor is, the lower the clock frequency will be. In this application, you will need a fairly low frequency, so I specify 25-μF capacitor in the parts list. You can experiment with other capacitor values for C1.

In operation, potentiometer R2 serves as a speed control for the sequence because this component controls the clock frequency. In some applications, manual control of the sequence speed might not be necessary. In such cases, simply substitute a fixed resistor with the appropriate value in place of the potentiometer (R2).

IC2 is a CD4013 dual flip-flop. This chip functions as a simple two-stage, four-step counter. The counter has two digital outputs. Because digital circuitry uses the binary system, each

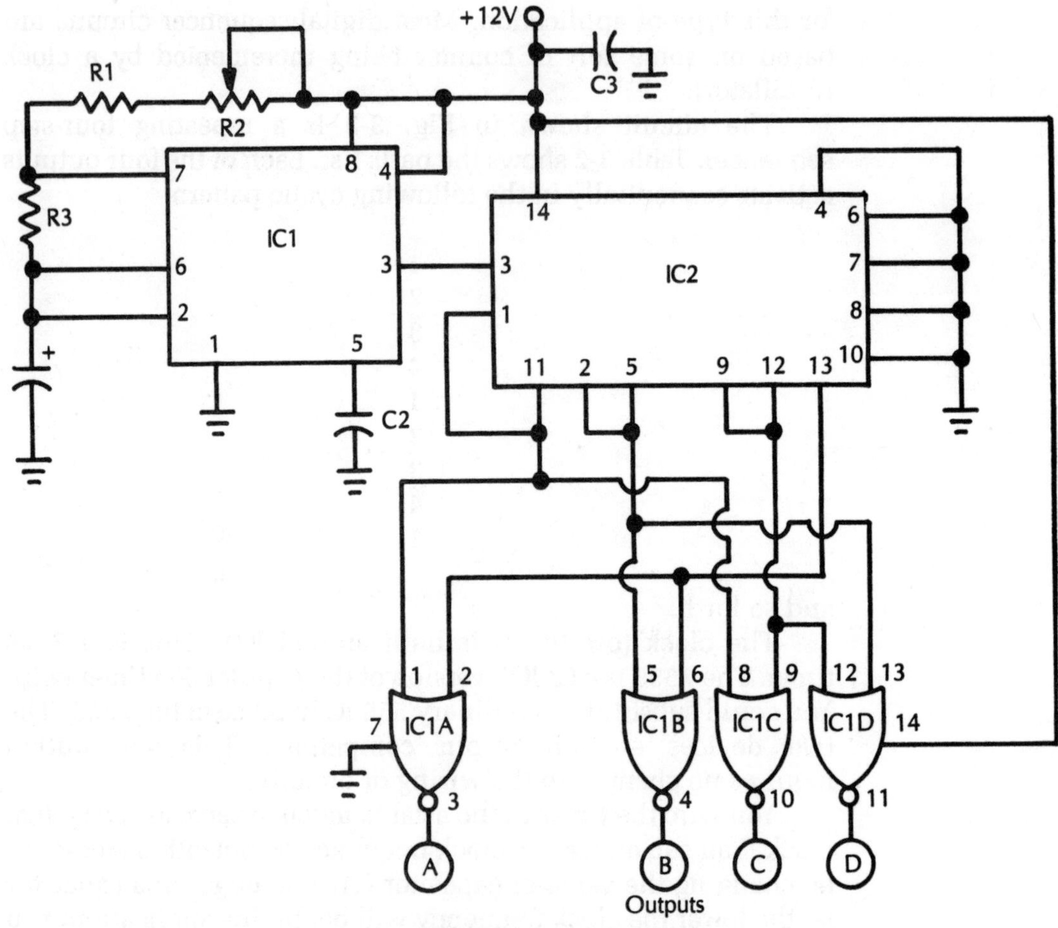

Fig. 3-3 *Project 9: Four-Step Sequencer.*

Table 3-2 Parts List for the Four-Step Sequencer of Project 9.

Part	Component
IC1	7555 timer (or 555)
IC2	CD4013 dual flip-flop
IC3	CD4011 quad NOR gate
C1	25-μF, 20-volt electrolytic capacitor (see text)
C2, C3	0.01-μF capacitor
R1, R3	1.2k, 1/4-watt resistor
R2	500k potentiometer (see text)

output can have two possible states. This system produces a total of four possible combinations:

A	B
0	0
0	1
1	0
1	1

The four NOR gates of IC3 convert the counter output into a one-of-four output pattern. There are four outputs from the circuit. Only one of the four outputs is active (high) at any given instant. The other three outputs are low.

Each of the output pulses is the same length. Anything you can control with a digital signal you also can control with this four-step sequencer project.

PROJECT 10: LONG-TIME SEQUENTIAL TIMER

Ordinary timer (monostable multivibrator) circuits are somewhat limited. There is a practical limit to the maximum time delay that can be achieved. For most circuits, the maximum time period is several minutes, or an hour or so at most.

If you need longer time periods, one solution is to use the XR2240 programmable timer. Before getting into this project, let me warn you that this IC might be a little difficult to locate. I recommend you try the surplus parts houses. Buy several if you can. This chip is very useful for the electronics experimenter.

Don't be put off by the term "programmable." The XR2240 is remarkably easy to use. In fact, eight of its sixteen pins are signal outputs. Selecting the desired output (or outputs) is all that is involved in "programming" this device. Figure 3-4 shows the pinout diagram for the XR2240 programmable timer.

This chip's internal circuitry consists essentially of a basic timer (not unlike the popular 555), followed by an eight-bit binary counter that provides the multiple outputs. A block diagram of the XR2240 programmable timer appears in Fig. 3-5.

The formula for the XR2240's basic time constant is different, however. Two external components, a resistor and a capacitor, control the base time period:

$$t = RC$$

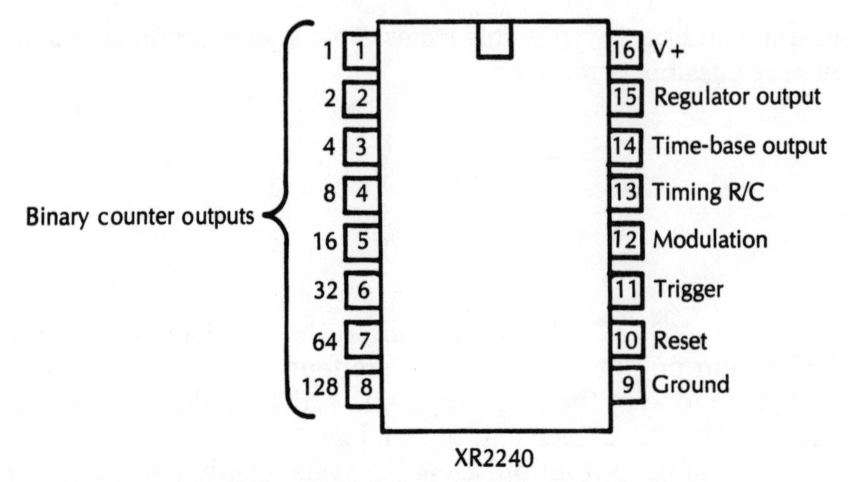

Fig. 3-4 *The XR2240 programmable timer IC is at the heart of the Long-Time Sequential-Timer Project.*

Unlike the 555 timer, the equation for the XR2240 programmable timer does not require a constant.

By using a single resistor/capacitor combination, you can use the XR2240 to provide 255 different output time periods. Each of these is an integer multiple of the base time period set by the values of resistor and capacitor according to the above equation.

The multiple outputs on the chip increase the base time period by binary intervals:

Pin #	Time Period
1	1t
2	2t
3	4t
4	8t
5	16t
6	32t
7	64t
8	128t

Each successive output pin simply doubles the time period of its immediate predecessor.

The operation of the XR2240 programmable timer is easiest to understand through an example. Let's assume that resistor R is

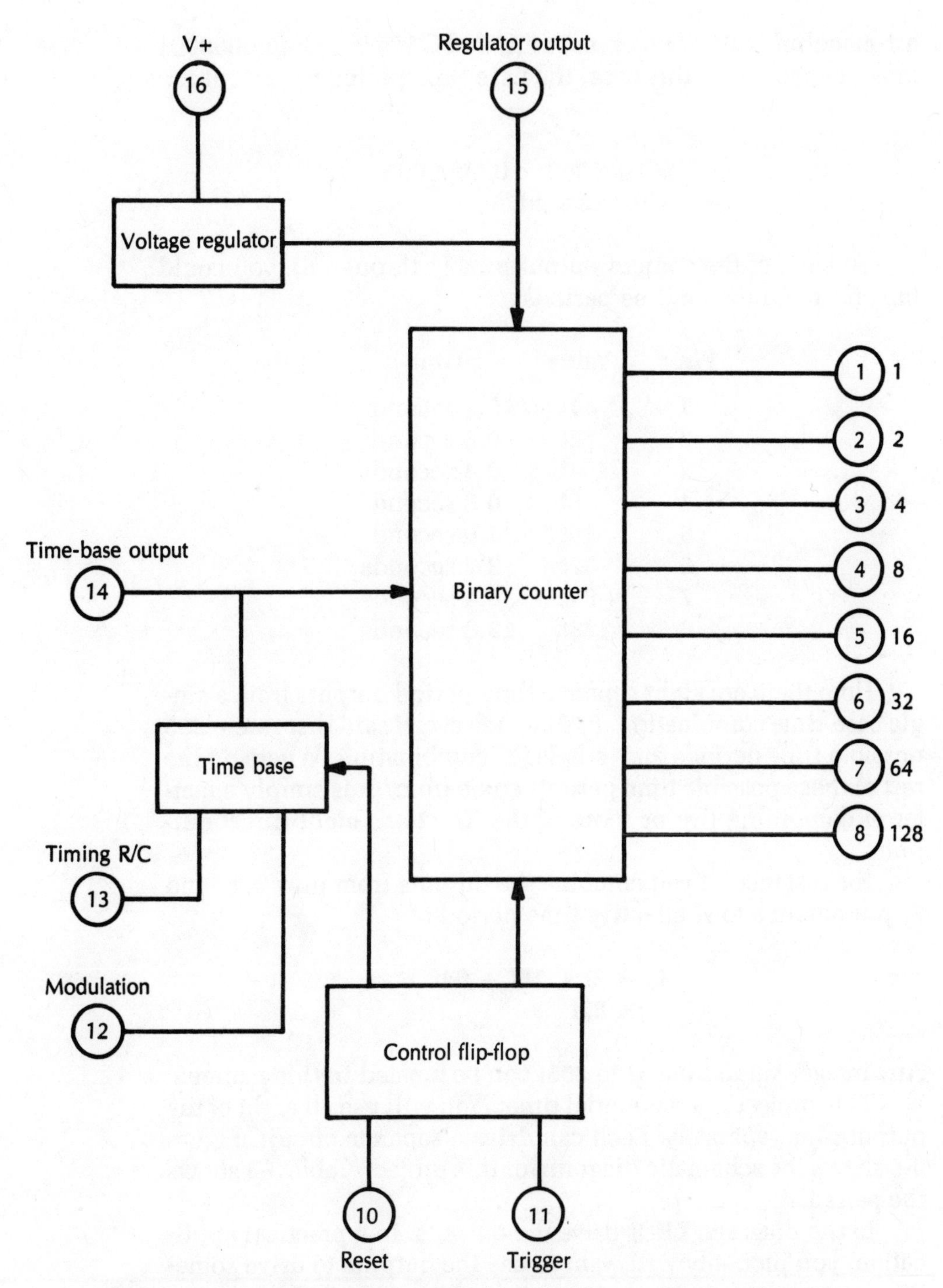

Fig. 3-5 *Block diagram of the XR2240 programmable timer IC's internal circuitry.*

a 1-megohm (1,000,000 ohms) unit, and C is a 0.1-μF (0.0000001 farad) capacitor. In this case, the base time period works out to:

$$
\begin{aligned}
t &= RC \\
&= 1000000 \times 0.0000001 \\
&= 0.1 \text{ second}
\end{aligned}
$$

At each of the counter output pins (1 through 8), you could tap off the following time periods:

Pin #	Value	Time
1	1t	0.1 second
2	2t	0.2 second
3	4t	0.4 second
4	8t	0.8 second
5	16t	1.6 second
6	32t	3.2 seconds
7	64t	6.4 seconds
8	128t	12.8 seconds

Here there are eight separate time-period outputs from a single base-time combination. Earlier, however, I said there were 255 possible time periods for a single RC combination. Where do the rest of these possible time periods come from? It is simply a matter of combining two or more of the XR2240's eight direct outputs.

For instance, if you combine the outputs from pins 2, 5, and 7, you obtain a total effective time period of:

$$
\begin{aligned}
t_x &= 2t + 16t + 64t \\
&= 82t
\end{aligned}
$$

Any integer value from 1t to 255t can be created in this manner.

This project is a sequential timer. You will use all eight of the output pins separately. Each can drive a separate circuit. Figure 3-6 shows the schematic diagram for this project. Table 3-3 shows the parts list.

In the diagram, LEDs drive the outputs. In a practical application, you probably will want to use the outputs to drive something more than a simple indicating device. In that case, you

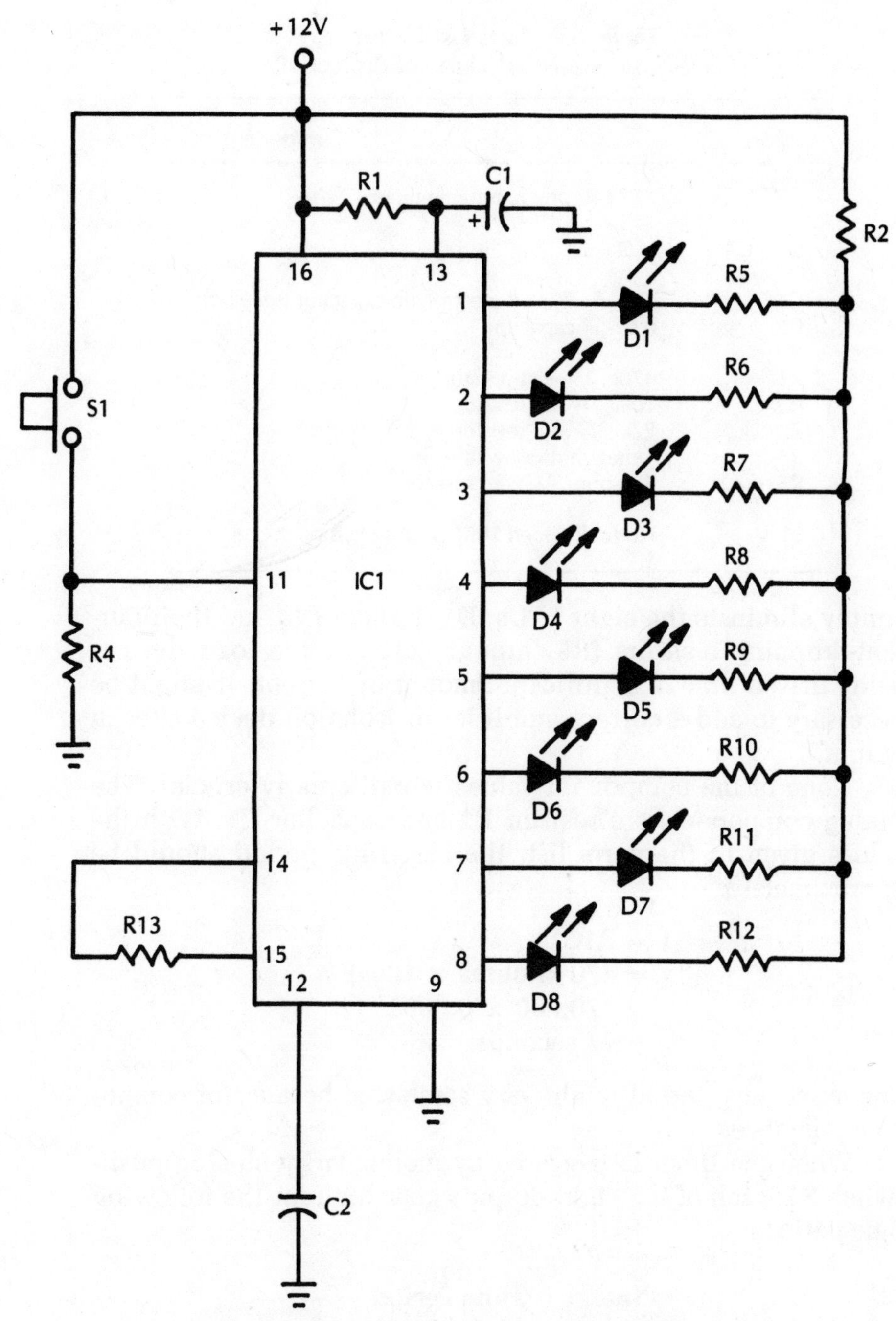

Fig. 3-6 Project 10: Long-Time Sequential Timer.

Table 3-3 Parts List for the Long-Time Sequential Timer of Project 10.

Part	Component
IC1	XR2240 programmable timer
D1 – D4	LED
C1	100-μF, 20-volt electrolytic capacitor (see text)
C2	0.01-μF capacitor
R1	470k, 1/4-watt resistor
R2	10k, 1/4-watt resistor
R3	22k, 1/4-watt resistor
R4	1-megohm, 1/4-watt resistor
R5 – R12	470-ohm, 1/4-watt resistor
S1	Normally open SPST push-switch

simply eliminate the eight LEDs (D1 through D8) and their current-dropping resistors (R5 through R12). If the load devices being driven draw a significant amount of current, it might be necessary to add a current amplifier or isolation device at each output.

None of the component values is particularly crucial. The timing components are resistor R1 and capacitor C1. With the values given in the parts list, the base-time period should be approximately:

$$\begin{aligned} t &= R1C1 \\ &= 470 \text{ kilohms} \times 100\mu F \\ &= 470{,}000 \times 0.0001 \\ &= 47 \text{ seconds} \end{aligned}$$

The exact time period might vary somewhat because of component tolerances.

When the timer is triggered by momentarily closing push-switch S1, each of the eight outputs goes high for the following time periods:

Pin #	Time Period
1	47 seconds
2	94 seconds

Pin #	Time Period
3	188 seconds
4	376 seconds
5	752 seconds
6	1504 seconds
7	3008 seconds
8	6016 seconds

Because of the long times involved, it might be more convenient to list the time periods in minutes rather than seconds:

Pin #	Time Period
1	0.78 minutes
2	1.57 minutes
3	3.13 minutes
4	6.27 minutes
5	12.53 minutes
6	25.07 minutes
7	50.13 minutes
8	100.27 minutes

It takes over an hour and a half for pin 8 to time out.

Of course, you can experiment with other values for the timing resistor (R1) and timing capacitor (C1).

PROJECT 11: TIMED TOUCH SWITCH

Touch switches are enjoyable and fascinating projects. They also come in very handy in certain circumstances. You can turn on any electrical device with just the lightest touch.

WARNING: because the user directly touches an exposed conductor, it is absolutely vital that any touch-switch project be operated on *battery power only*! Do not use an ac-to-dc converter to power this type of circuit. *Never* use *any* form of ac power with *any* touch switch! No matter how careful you are, accidents can occur. If there is the right kind of short anywhere in the circuit, the result could be an extremely painful, if not fatal, shock. It is not worth the risk, no matter how unlikely you think an accident might be. Take all reasonable precautions.

A practical touch-switch circuit appears in Fig. 3-7. Table 3-4 shows the parts list. This circuit also includes a timer. The output of the circuit is normally low. When you short the two touch

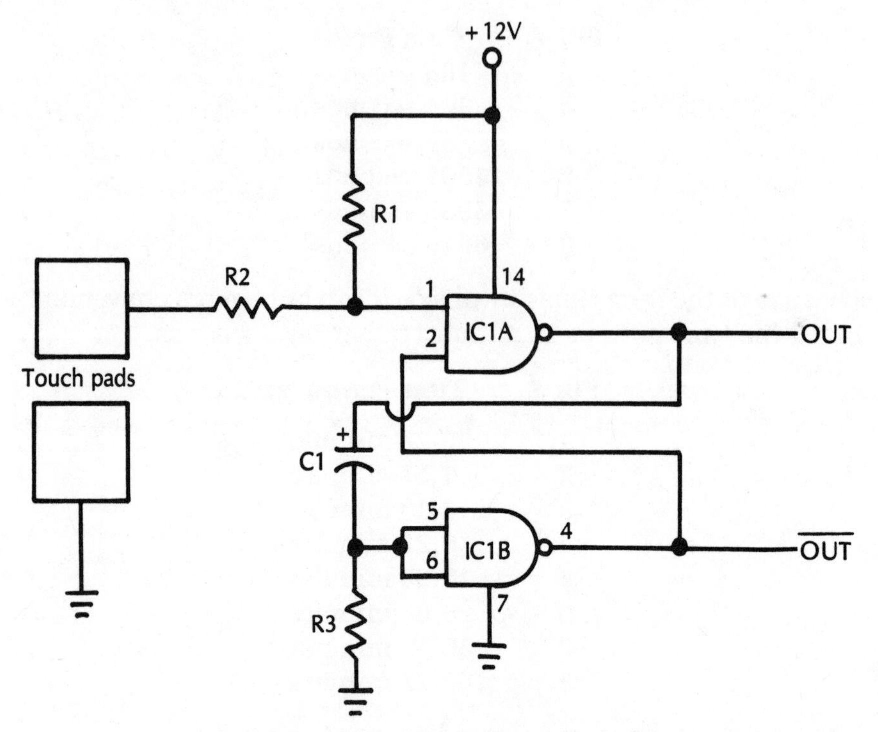

Fig. 3-7 Project 11: Timed Touch Switch.

Table 3-4 Parts List for the Timed Touch Switch of Project 11.

Part	Component
IC1	CD4011 quad NAND gate
C1	5-μF, 20-volt electrolytic capacitor*
R1	10-megohm, 1/4-watt resistor
R2, R3	220k, 1/4-watt resistor*

Important! Use battery power only!

*Timing component—see text

pads with your fingertip (or almost anything else), the output goes high for a specific period of time, and then goes low again. The length of time the touch pads are shorted together is irrelevant. The output automatically goes low after the preset time period, even if the touch pads are shorted continuously.

The time period is set by the values of capacitor C1 and resistor R3. With the component values called for in the parts list, the on time is approximately one second. You can experiment with other part values.

Before moving on to the next project, the safety warning bears repeating: any touch-switch project should be operated on *battery power only*! *Never* use *any* form of ac power with *any* touch switch! Please don't take foolish chances.

If you want to operate an ac-powered electrical device from a touch switch, use the touch-switch circuitry to drive a relay or some other isolating device.

PROJECT 12: ONE-SECOND TIMER

Sometimes you need to time something, but you can't watch a clock and watch for the event simultaneously. What is needed in such cases is an audible timekeeping device. A circuit for this purpose appears in Fig. 3-8. Table 3-5 shows the parts list.

This project puts out a tone or beep once per second. Depending on the application, you might or might not need to count the beeps. In any case, your eyes can be occupied elsewhere.

Much of this circuit is a 1-Hz (one pulse per second) time base. This portion of the circuit also occurs in the Digital Clock and Countdown Timer projects that are presented also in this chapter.

The manufacturer recommends the following values for use with the MM5369 60 Hz time-base chip (IC1):

C1 6.36pF
C2 30pF

It probably will be difficult to find capacitors with these values. Fortunately, this chip is not too fussy. The following capacitor values should be close enough for all but the most critical timing applications:

C1 10pF
C2 47pF

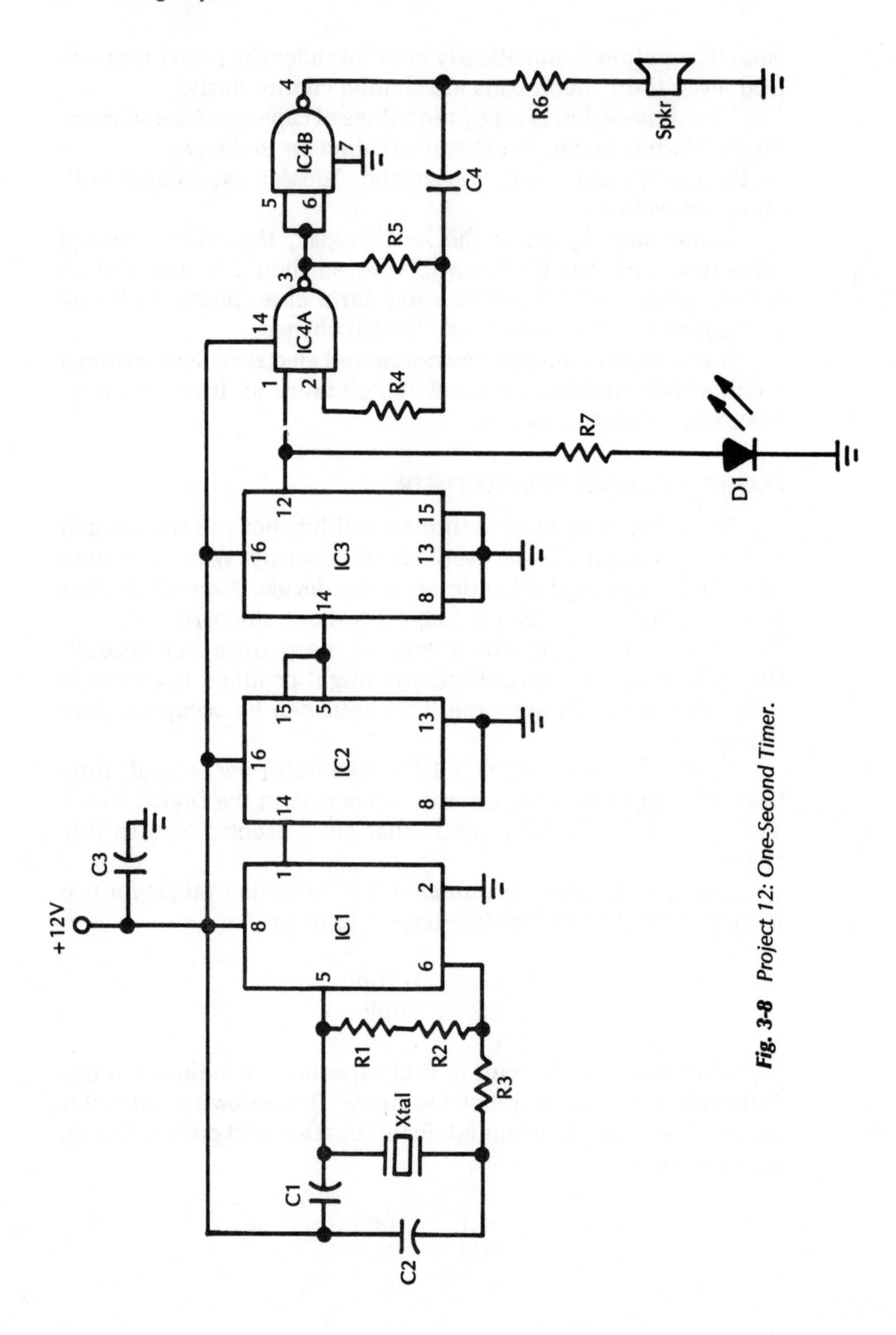

***Fig. 3-8** Project 12: One-Second Timer.*

Table 3-5 · Parts List for the One-Second Timer of Project 12.

Part	Component
IC1	MM5369 60-Hz time base
IC2, IC3	CD4017 decade counter
IC4	CD4011 quad NAND gate
D1	LED
C1	10-pF (6.36pF) capacitor
C2	47-pF (30pF) capacitor
C3	0.1-μF capacitor
C4	0.01-μF capacitor
R1, R2	10-megohm, 1/4-watt resistor
R3	1k, 1/4-watt resistor
R4	1-megohm, 1/4-watt resistor
R5	100k, 1/4-watt resistor
R6	100-ohm, 1/4-watt resistor
R7	390-ohm, 1/4-watt resistor
Spkr	Small speaker
Xtal	3.58-MHz color-burst crystal

The MM5369 converts the 3.58-MHz crystal frequency into a 60-Hz signal. Two counter stages (IC2 and IC3) reduce this frequency further to 1 Hz.

The output pulses drive a gated audio oscillator made up of two sections of a quad NAND gate package (IC4). The tone sounds only when the signal fed to pin 1 (the output from the 1-Hz time-base circuitry) goes high. The result is one beep from the speaker for each pulse. Because there is one pulse per second, the speaker emits a burst of tone once per second.

You can adjust the output volume by changing the value of resistor R6. Altering the value of resistor R5 changes the frequency of the tone. You can replace either or both of these fixed resistors with a potentiometer if you desire. Capacitor C4's value also affects the tone frequency.

For additional convenience, the 1-Hz signal also is displayed on an LED (D1) that flashes once per second.

PROJECT 13: COUNTDOWN TIMER

In some applications, it can be quite useful to have a time-keeping device that counts backwards. This project is a single-digit countdown timer. The project can be expanded easily to incorporate additional stages if you choose. To keep the schematic diagram as clear as possible, it is broken into two sections that are shown in Figs. 3-9 and 3-10. Table 3-6 shows the parts list.

Figure 3-9 shows the manual switching network to set the starting value. You can enter any value from 1 to 9. The switching network ensures that the input is in proper BCD format.

The actual countdown circuit is shown in Fig. 3-10. The input signal to be counted is derived from a 1-Hz time base (IC2, IC3, and IC4, and their associated components). This circuit also occurs in the Digital Clock and One-Second Beeper projects presented elsewhere in this chapter.

The manufacturer recommends the following values for use with the MM5369 60-Hz time-base chip (IC2):

C1	6.36pF
C2	30pF

It probably will be difficult to find capacitors with these values. Fortunately, this chip is not too fussy. The following capacitor values should be close enough for all but the most critical timing applications:

C1	10pF
C2	47pF

The MM5369 converts the 3.58-MHz crystal frequency into a 60-Hz signal. Two counter stages (IC3 and IC4) reduce this frequency further to 1 Hz. You can substitute an external time base for this portion of the circuit if you prefer.

Using a 1-Hz time base, the countdown timer can count from one to nine seconds. By adding two more decade counters (set up to divide by six and ten, respectively), you can extend the countdown time range from one to nine minutes. This modification can be made easily to the circuit; just duplicate both IC3 and IC4, using the same wiring shown for these components. Of course, if

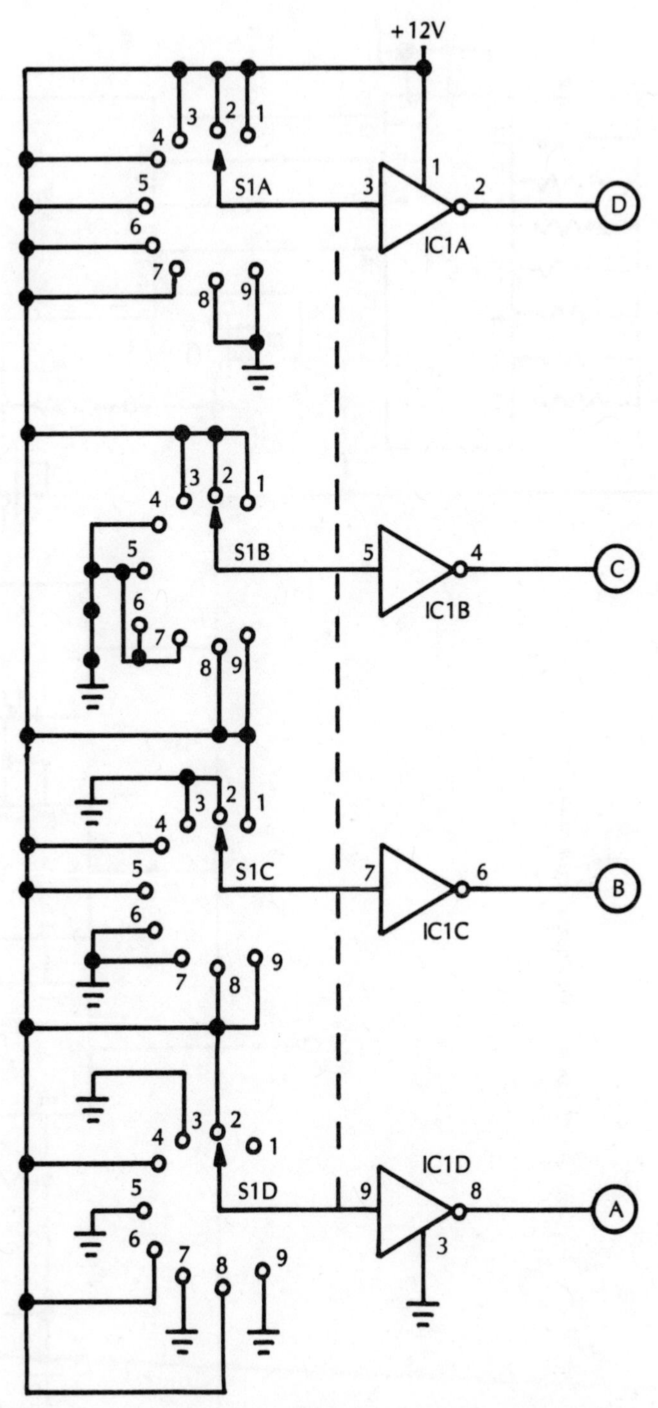

Fig. 3-9 *Project 13: Manual switching network for Countdown Timer project.*

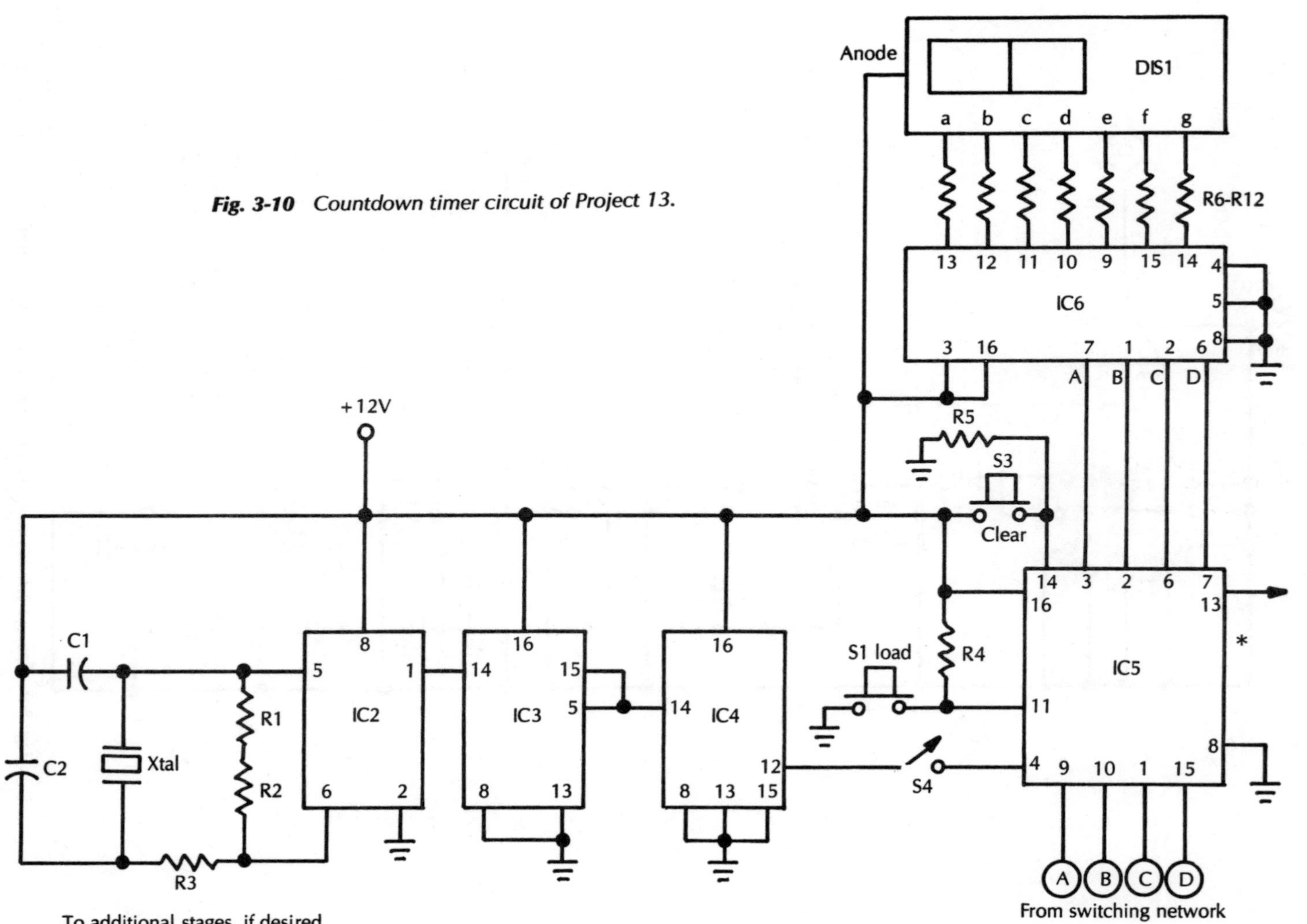

Fig. 3-10 *Countdown timer circuit of Project 13.*

Table 3-6 Parts List for the Countdown Timer of Project 13.

Part	Component
IC1	CD4049 hex inverter
IC2	MM5369 60-Hz time base
IC3, IC4	CD4017 decade counter
IC5	74C193 up/down counter
IC6	74C46 BCD-to-seven-segment-display decoder/driver
DIS1	Seven-segment LED display, common anode
C1	10-pF (6.36pF) capacitor
C2	47-pF (30pF) capacitor
R1, R2	10-megohm, 1/4-watt resistor
R3 – R5	1k, 1/4-watt resistor
R6 – R12	330-ohm, 1/4-watt resistor
S1	4-pole, 9-throw rotary switch
S2, S3	SPST pushbutton switch—normally open
S4	SPST switch
Xtal	3.58-MHz color-burst crystal

you use some other time base, the countdown units will vary accordingly.

Another possible modification to this project is to include some additional counting stages for multidigit counts. A two-stage circuit could count down for a starting value as high as 99. Just duplicate everything in the circuit shown here (both Figs. 3-9 and 3-10), except the components associated with the time-base circuit. The second stage is driven by the "borrow out" signal from the 74C193 up/down counter (IC5). This tap-off point is marked "*" in Fig. 3-10.

PROJECT 14: CAR CLOCK

Most of us have fairly busy schedules and generally need to keep track of the time. A car clock can be nice to have, especially for people who spend quite a bit of time on the road.

Unfortunately, many cars do not have clocks. Those that do often have mechanical clocks that have a reputation for being unreliable. I've known countless people who have had a dead clock in their car. If it ran at all, half the time it was so inaccurate that it was virtually useless.

A digital clock is fairly easy to build, isn't very expensive, and is usually quite reliable. Because the clock must keep running at all times, its power connection to the car battery must come before the ignition switch. To minimize drain on the battery, you could use two separate power connections, one for the timekeeping circuitry, and the other for the display LEDs. The timekeeping circuitry should run continuously. The LED power line can come after the ignition switch, so the display lights up only when the ignition is on.

Although I present this project here as a car clock, it can be run off of any 12-volt dc power source.

Digital clocks usually are comprised of an accurate (generally crystal controlled) time base and a series of counter stages. The time base is an oscillator with a very precisely set frequency. The counter stages keep track of how many oscillator cycles have gone by. If you know the frequency and the number of time-base pulses, you can determine the time accurately. Of course, this is all done automatically.

A convenient time-base frequency for a clock is 1 Hz, or one pulse per second. Sixty pulses equal one minute. Sixty minutes (60 × 60 pulses) equals one hour. An accurate 1-Hz oscillator isn't very convenient, so 60-Hz oscillators normally are used, and additional counter stages are used to reduce the 60-Hz signal to 1 Hz. There are 60 pulses per second.

For a clock to be useful, the time-base frequency must be extremely accurate. It might seem that there is very little difference between 0.95 second and 1.00 second. However, when you multiply that 5 percent error over a 24-hour day, you end up with a 22-hour, 48-minute day. That's not very good timekeeping. A clock that inaccurate would be useful only as a paperweight.

Most digital clocks work with some sort of crystal oscillator. Quartz crystals make very precise frequencies possible. Incidentally, when digital watches first became popular, many (especially the more expensive models) were touted proudly in the ads as being quartz controlled. Actually, all digital watches are quartz controlled. A crystal oscillator is used for the time base, and crystals are merely slabs of quartz.

Most crystals operate at frequencies much higher than 60 Hz. Generally, their resonant frequencies are above 1 MHz (1,000,000 Hz). Additional counter stages are needed to drop this high frequency down to where you want it.

The MM5369 (shown in Fig. 3-11) is an IC designed especially for this purpose. It generates an extremely precise and stable 60-Hz time-base signal from a 3.579545-MHz (3,579,545 Hz) crystal. This odd frequency usually is rounded off to 3.58 MHz for convenience in talking or writing about it. This particular input frequency was selected because it is the frequency used in color-burst oscillators in color TV sets. Suitable 3.58-MHz crystals are widely available. It is probably the most commonly used crystal frequency.

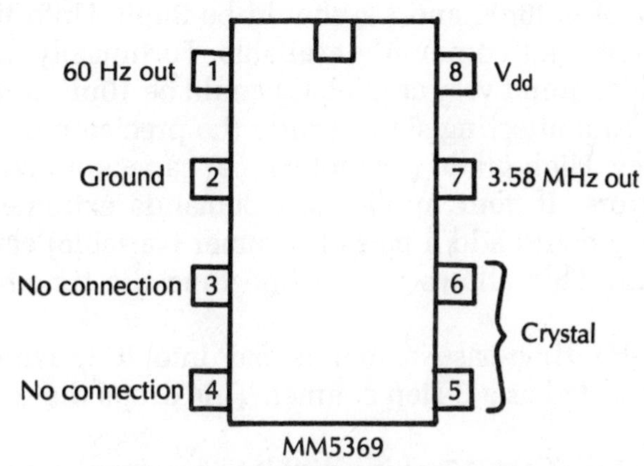

Fig. 3-11 The MM5369 time-base IC simplifies the construction of digital clock circuits.

In the MM5369, a single 8-pin chip provides all the necessary frequency division. Actually, except for the power connections (V_{dd} and ground), you normally use only three pins. One more pin is optional. The remaining two pins serve no electrical purpose. They are only there to make up a standard-sized IC DIP package.

Two of the active pins are connected to either end of the 3.58-MHz crystal. The remaining pin is the output, providing the 60-Hz time-base signal. In case it might be needed in some circuits, pin 7 also provides a 3.58-MHz signal. Pins 3 and 4 are not connected internally within the chip.

Because this project is a little more complex than most of the others in this book, I have broken up the schematic into three main sections. These sections can be identified as the seconds counter, minutes counter, and hours counter. These sections are shown and discussed separately.

Seconds Section

First you must divide the 60-Hz time-base signal by 60 to get a 1-Hz signal that can be counted by the clock. Figure 3-12 shows how the MM5369 is used to create a 1-Hz output signal. Note that three resistors and two capacitors occur in the circuit with the crystal itself. These components improve the stability of the circuit. According to the manufacturer of the MM5369, C1 should have a value of 6.36pF, and C2 should be 30pF. Unfortunately, these values are not commonly available. Fortunately, however, the exact values aren't very crucial. C1 could be 10pF, and C2 can be 47pF without affecting significantly the precision of the output frequency. High-grade (low tolerance) capacitors will minimize the errors. If your application demands extremely high precision, you could add a pair of trimmer (variable) capacitors in the circuit. This allows you to fine tune the time-base frequency.

The 60-Hz time-base signal is fed into IC2, which is a CD4017 connected as a 6-step counter. This drops the frequency to 10 Hz.

IC3, another CD4017 chip, divides the signal by an additional factor of ten, producing a precise 1-Hz output.

Minutes Section

Connecting the 1-Hz signal-source circuit of Fig. 3-12 to the input of the circuit shown in Fig. 3-13 provides a clock with minute readouts of 00 to 59. Once the count reaches 60, the counters (and display) are reset to 00. The count, or current minute number, is displayed on a pair of seven-segment LED displays.

Note that this circuit also features a 1-minute output. This signal is used in the hours counter section presented next.

Hours Section

Figure 3-14 shows the circuit for adding the display for hours, ranging from 01 to 12. After the count exceeds 12, the

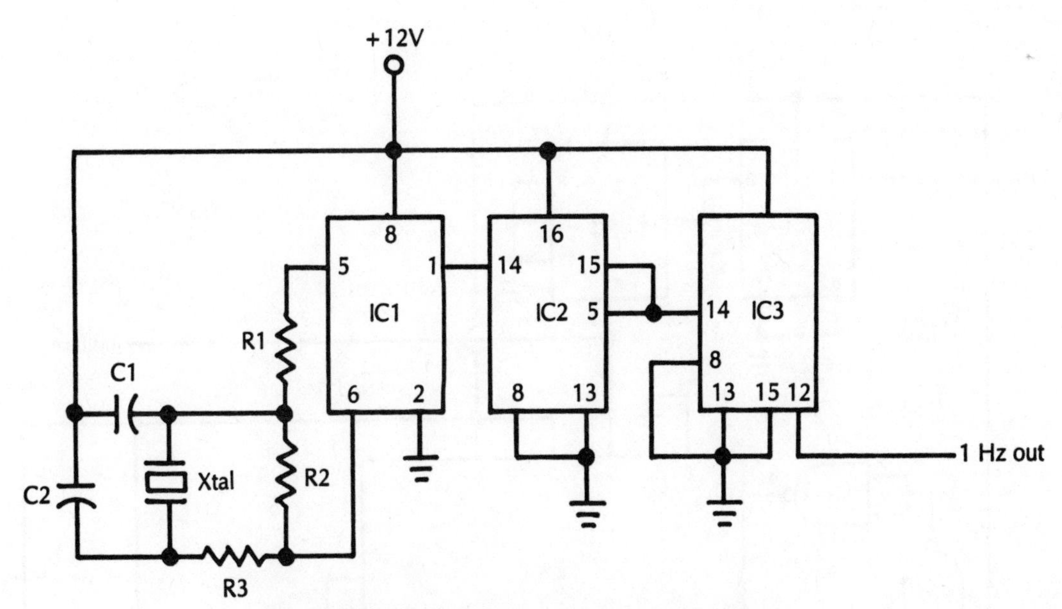

Fig. 3-12 Seconds section of Project 14: Car Clock.

counters and display are reset to 01, and the count starts over. IC10 and IC11 count each group of 60 minutes (one hour).

Because the hours display always reads either 0 or 1 in the tens position, you wire this display unit a little differently. You wire the ones digit in the usual fashion, but you can use a few short-cuts for the tens digit. This digit is always either 0 or 1. Segments b and c are always lit, so they are tied directly to the positive power supply (through current-dropping resistors, of course). If the count is less than 10 (1 through 9), then segments a, d, e, and f also light to produce a 0. Segment g never lights and is left disconnected.

A block diagram of the entire car clock project appears in Fig. 3-15. Table 3-7 gives the complete parts list for this project.

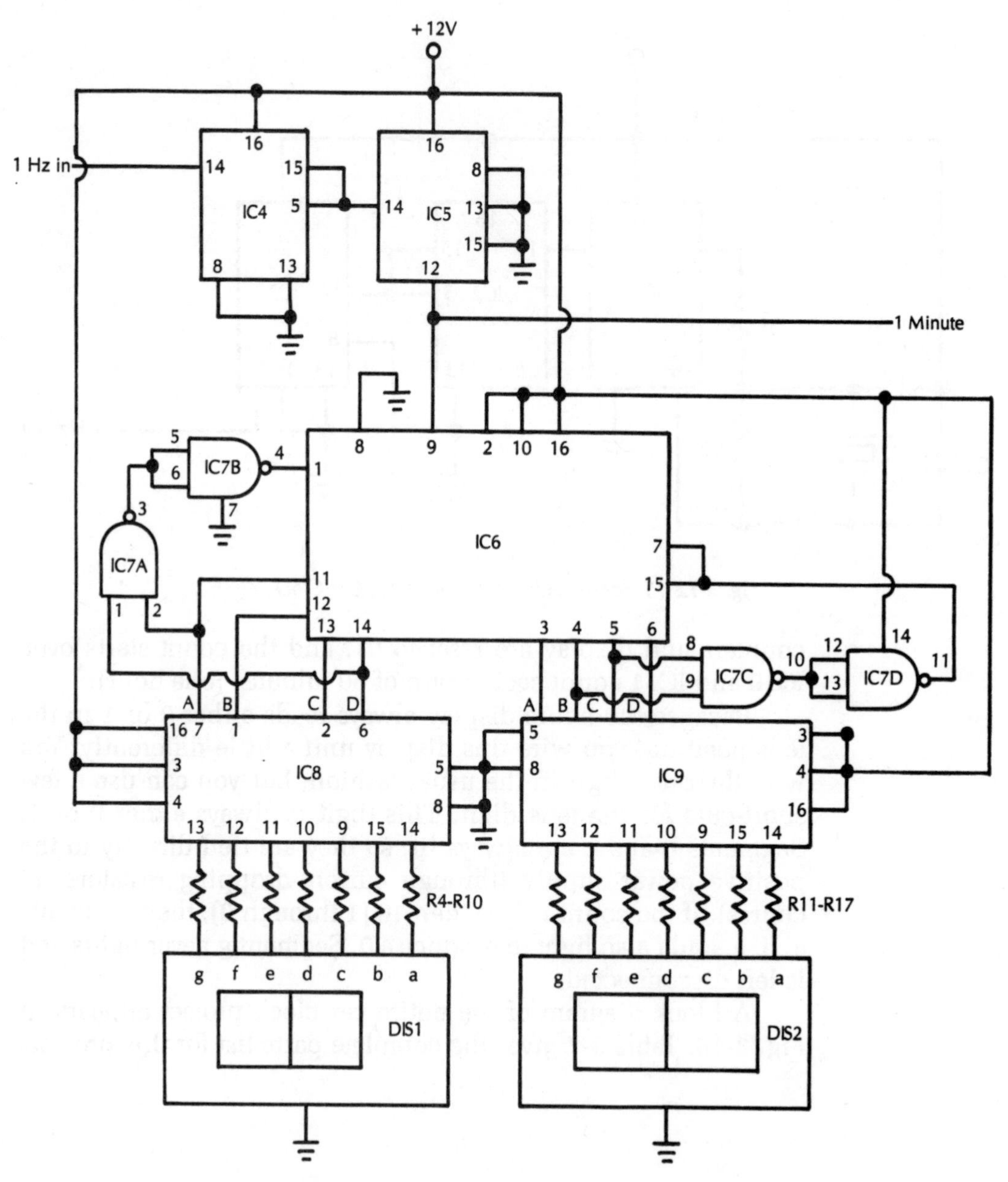

Fig. 3-13 Minutes section of the Car-Clock project.

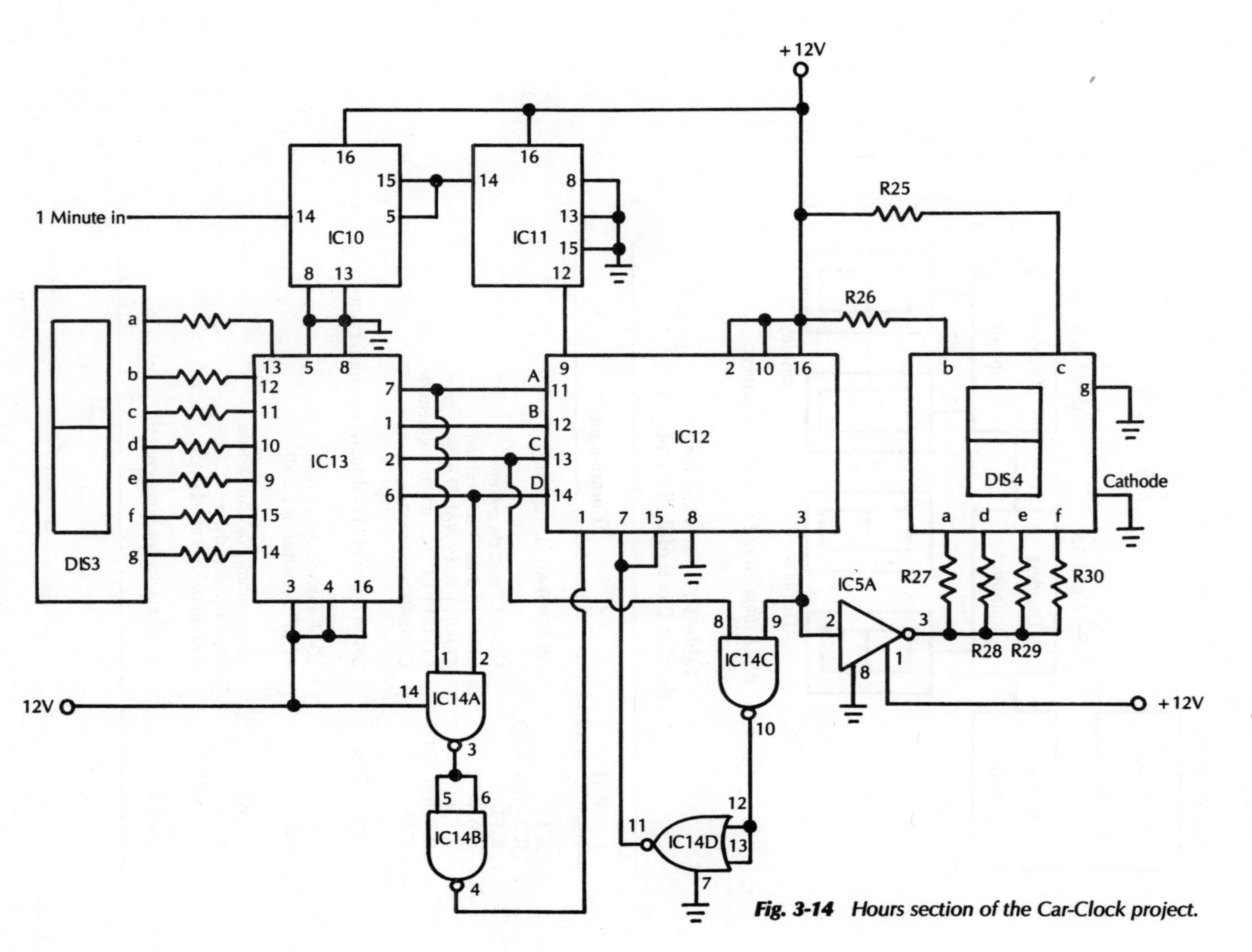

Fig. 3-14 Hours section of the Car-Clock project.

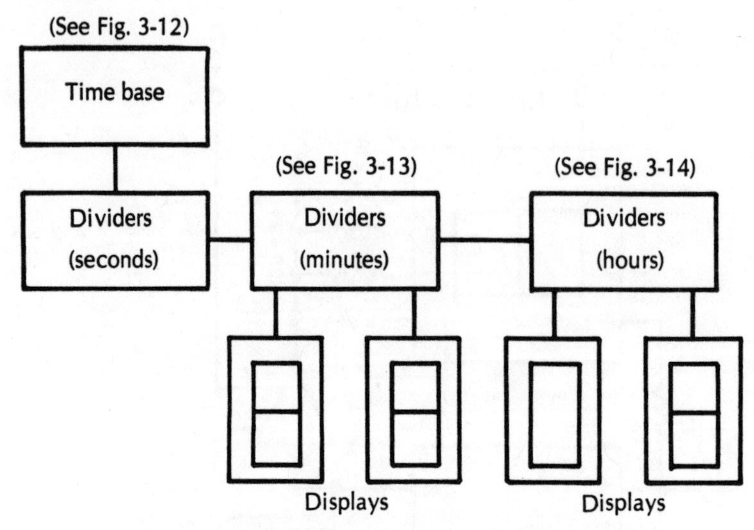

Fig. 3-15 *Block diagram of the Car-Clock project.*

Table 3-7 Parts List for the Car Clock of Project 14.

Part	Component
IC1	MM5369 60 Hz time base
IC2 – IC5, IC10, IC11	CD4017 decade counter
IC6, IC12	CD4518 dual BCD counter
IC7, IC14	CD4011 quad NAND gate
IC8, IC9, IC13	CD4511 BCD-to-seven-segment decoder
IC15	CD4049 hex inverter
DIS1 – DIS4	Seven-segment LED display, common cathode
C1	10-pF capacitor (see text)
C2	47-pF capacitor (see text)
R1, R2	10-megohm, 1/4-watt resistor
R3	1k, 1/4-watt resistor
R4 – R30	330-ohm, 1/4-watt resistor
Xtal1	3.58-MHz color-burst crystal

❖4 Communications Projects

EARLY ELECTRONICS HOBBYISTS WORKED MAINLY WITH RADIO projects, and communications are still a big part of the field.

This chapter describes several projects that you can incorporate into a communications system. Some are of interest primarily to ham-radio operators, while others have a more general appeal. This chapter concludes with a pair of related projects—a complete optical communications system (transmitter and receiver). With these two projects, you will be able to send messages from one location to another via a beam of infrared light.

PROJECT 15: CPO

Ham-radio operators usually start out working with Morse Code. To learn this code of "dots" and "dashes" takes quite a bit of practice. This practicing can be done most conveniently with a *code practice oscillator*, or CPO. This circuit produces a tone when you open and close a switch according to Morse-Code patterns.

A simple CPO circuit appears in Fig. 4-1. Table 4-1 shows the parts list.

Nothing is crucial in this circuit. Almost any transistors can be used. Just remember that Q1 must be an npn type and Q2 should be a pnp transistor. You should select the two transistors for similar electrical characteristics.

Switch S1 ideally should be a paddle-type code key. If such a device is not available, you can substitute an ordinary normally open SPST pushbutton switch. The pushbutton should be as

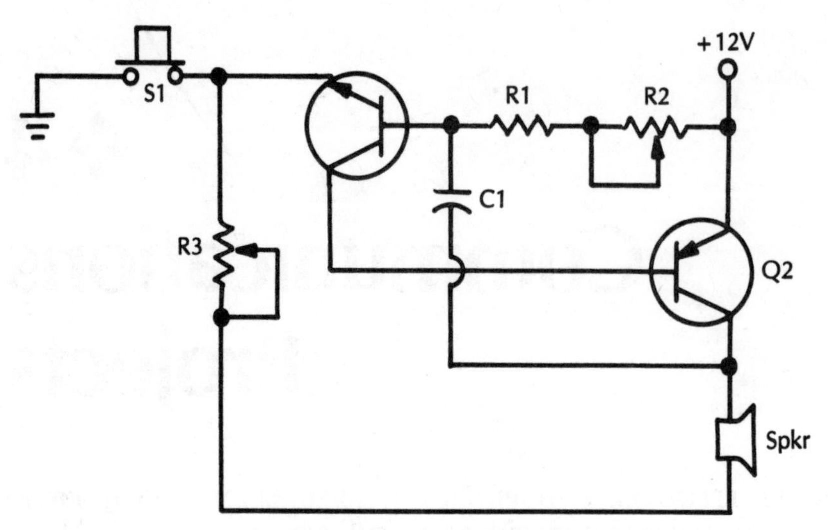

Fig. 4-1 Project 15: CPO.

Table 4-1 Parts List for the CPO of Project 15.

Part	Component
Q1	Npn transistor (2N3904, or similar)
Q2	Pnp transistor (2N3906, or similar)
C1	0.1-μF capacitor
R1	100k, 1/4-watt resistor
R2	500k potentiometer (pitch)
R3	500-ohm potentiometer (volume)
S1	Code-key switch, or normally open SPST pushbutton switch
Spkr	Small speaker

large as possible. It is very difficult to tap out Morse Code on a small button.

Potentiometer R2 adjusts the frequency or pitch of the tone produced by the speaker. Set this control for the most comfortable and pleasing sound. If you prefer, you could replace R1 and R2 with a single fixed resistor with a value between 100k (100,000 ohms) and 600k (600,000 ohms).

Potentiometer R3 permits you to reduce the volume of the tone. You can replace it with a fixed resistor or eliminate it altogether. The higher the value of R3, the lower the volume will be.

If you prefer, you could use a set of headphones in place of the speaker. If you use headphones, some attenuation of the volume is essential. At full volume, the tone probably will hurt your ears.

For your convenience, Table 4-2 gives a summary of the Morse Code.

Table 4-2 Morse Code.

Letter	Code Pattern
A	.-
B	-...
C	-.-.
D	-..
E	.
F	..-.
G	- -.
H	....
I	..
J	.- - -
K	-.-
L	.-..
M	- -
N	-.
O	- - -
P	.- -.
Q	- -.-
R	.-.
S	...
T	-
U	..-
V	...-
W	.- -
X	-..-
Y	-.- -
Z	- -..
0	- - - - -
1	.- - - -
2	..- - -
3	...- -
4	....-
5	.....
6	-....
7	- -...
8	- - -..
9	- - - -.

PROJECT 16: SCA DECODER

Many FM radio stations broadcast a special secondary signal primarily for use in businesses. This signal is known as SCA (Subsidiary Communication Authorization). Some SCA broadcasts consist of specialized announcements, but generally the SCA signal is used for background music.

With the circuit shown in Fig. 4-2, you can receive SCA signals in your own home. Table 4-3 shows the parts list. The input signal is taken from the output of the FM demodulator of a standard FM radio or receiver.

It is important to mention that this project is intended for private use only. Any commercial use of SCA signals without paying the required license fees is illegal.

The heart of this circuit is a PLL (phase-locked loop). The PLL is IC1 (567). The decoder is tuned by potentiometer R6. You might want to use a screwdriver-adjusted trimpot for this control, because it normally won't be readjusted during use. Tuning the SCA decoder is very easy. You only have to adjust R6 for approximate tuning. The nature of a PLL causes it to fine-tune itself. It will seek out and lock onto the SCA signal.

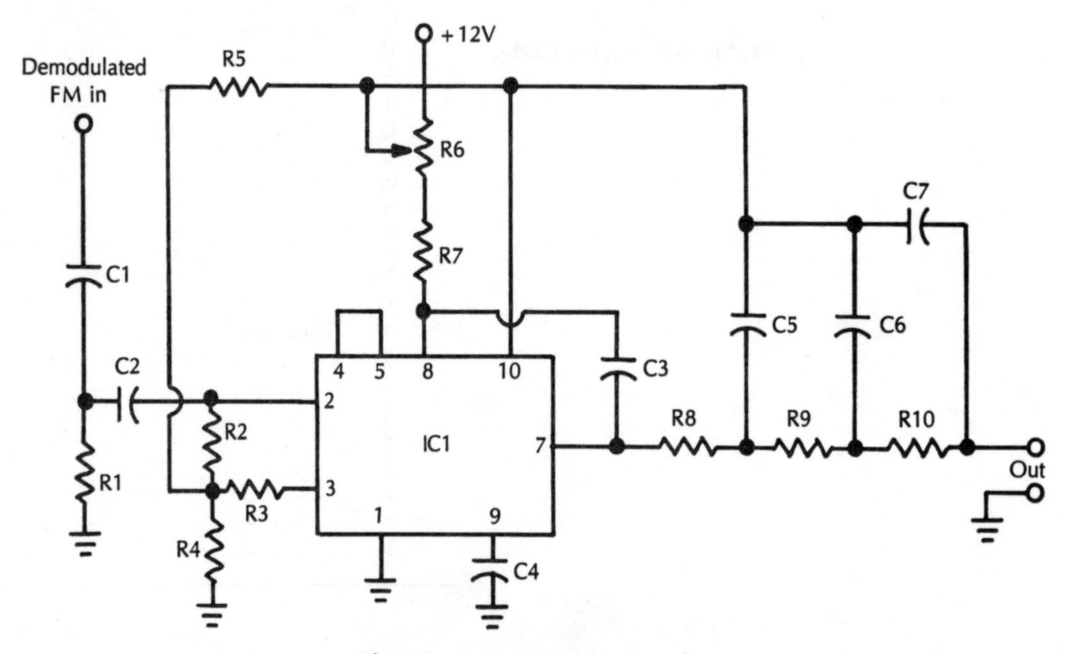

Fig. 4-2 Project 16: SCA Decoder.

Table 4-3 Parts List for the SCA Decoder of Project 16.

Part	Component
IC1	565 PLL
C1, C2	510-pF capacitor
C3, C4	0.001-μF capacitor
C5, C7	0.018-μF capacitor
C6	0.05-μF capacitor
R1 – R4	4.7k, 1/4-watt resistor
R5	10k, 1/4-watt resistor
R6	5k potentiometer (see text)
R7	1.8k, 1/4-watt resistor
R8 – R10	1k, 1/4-watt resistor

Resistors R8 through R10 and capacitors C5 through C7 are a filter network for deemphasis. In FM broadcasts (including SCA broadcasts), certain frequencies are emphasized, or *boosted*, before transmission. This is done for noise reduction. Frequencies likely to be lost in noise receive an extra boost. These boosted frequencies must be cut back (deemphasized) at the receiver to recreate the original audio signal.

The deemphasis network also reduces the audibility of high-frequency hiss. SCA signals usually tend to be quite hissy, so this filtering is highly desirable.

PROJECT 17: FM STEREO CONVERTER

Stereo (two-channel) sound can enhance many kinds of listening situations. Home hi-fi systems are now universally stereo. Almost all FM broadcast stations transmit in stereo, but there are still a lot of mono FM radios around. If you have one of these units, you might not be getting the maximum possible pleasure out of listening to music.

This project allows you to convert any mono FM receiver into stereo. Figure 4-3 shows the circuit. The parts list appears in Table 4-4.

The lamp (I1) lights up to indicate when you are receiving a stereophonic signal. You use potentiometer R3 to fine-tune the IC's internal VCO (voltage-controlled oscillator).

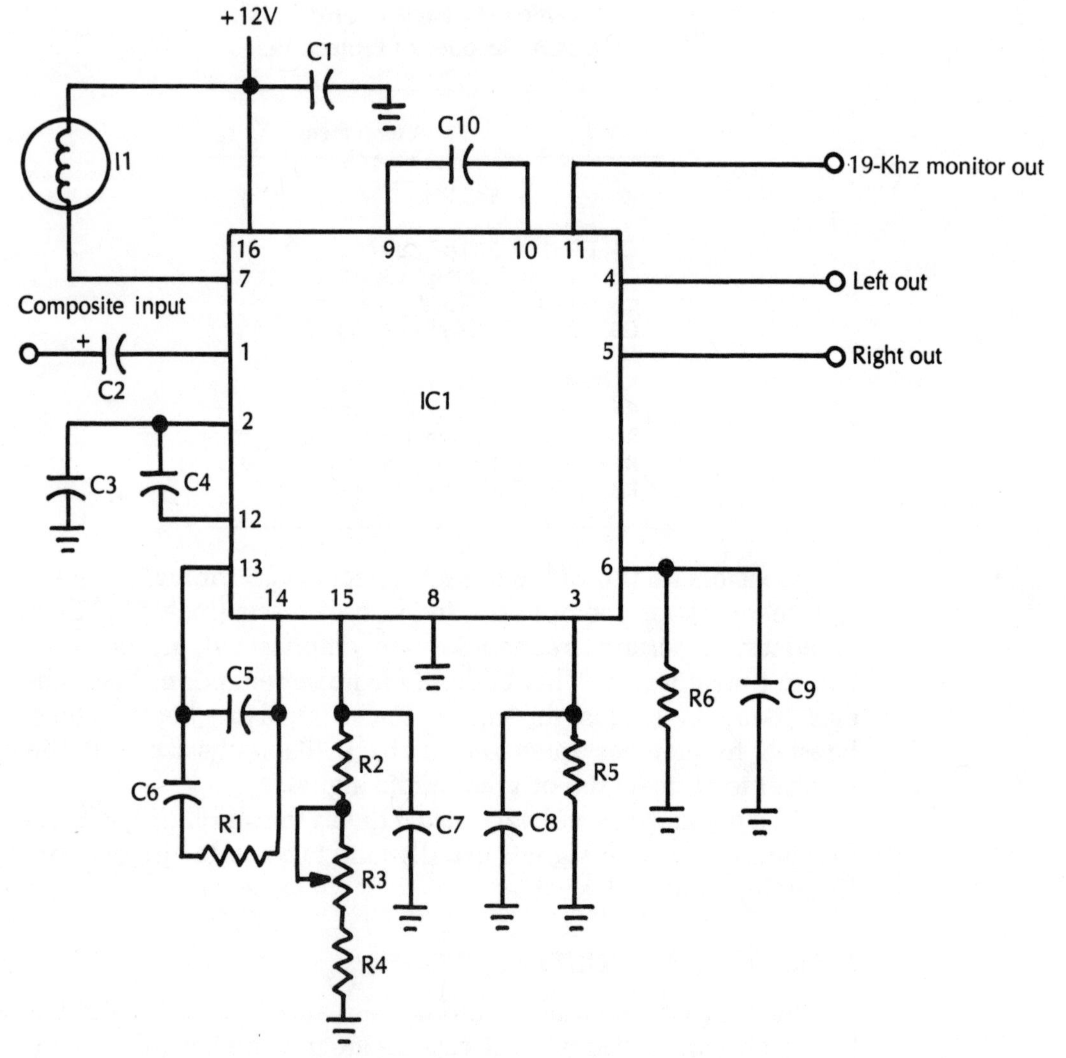

Fig. 4-3 *Project 17: FM Stereo Converter.*

PROJECT 18: AUTOMATIC TV SHUT-OFF

Have you ever fallen asleep with the TV on? You wake up in the wee hours to that obnoxious hissing sound and a screenful of snow because the station has gone off the air, and there is no signal for your set to receive. This wastes power and is annoying.

This project provides a convenient and automatic solution. The circuit appears in Fig. 4-4. Table 4-5 shows the parts list.

Table 4-4 Parts List for the FM Stereo Converter of Project 17.

Part	Component
IC1	LM1800
C1	0.1-μF capacitor
C2	2.2-μF, 20-volt electrolytic capacitor
C3	0.0022-μF capacitor
C4	0.0033-μF capacitor
C5	0.22-μF capacitor
C6	0.5-μF capacitor
C7	390-pF capacitor
C8, C9	0.022-μF capacitor
C10	0.33-μF capacitor
R1	3.3k, 1/4-watt resistor
R2	18k, 1/4-watt resistor
R3	5k potentiometer
R4	2.2k, 1/4-watt resistor
R5, R6	3.9k, 1/4-watt resistor
I1	12-volt lamp—100 mA

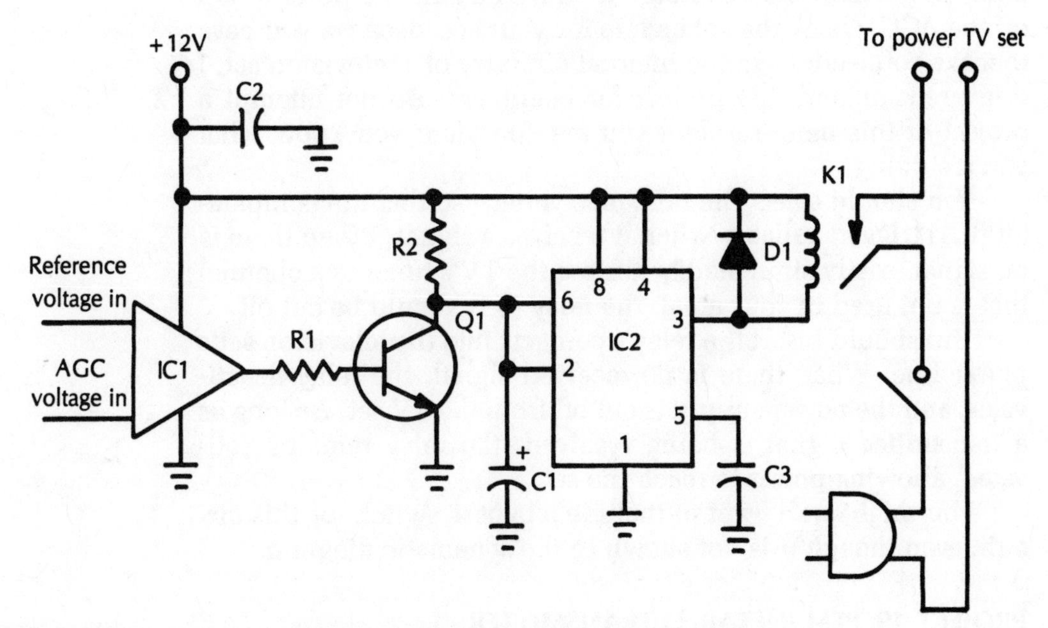

Fig. 4-4 Project 18: Automatic TV Shut-Off.

Table 4-5 Parts List for the Automatic TV Shut-Off of Project 18.

Part	Component
IC1	531
IC2	555
Q1	Npn transistor (2N3904, or similar)
D1	Diode (1N4003, or similar)
C1	10-μF, 25-volt electrolytic capacitor
C2	0.1-μF capacitor
C3	0.01-μF capacitor
R1	1k, ½-watt resistor
R2	1-megohm watt resistor
K1	Relay with contacts to suit load

In operation, the circuit requires two input signals. The AGC (automatic gain control) voltage is tapped off from an appropriate point in the television's circuitry. To find a suitable point to tap off the AGC, check the schematic for your set. Because you have to make connections to the internal circuitry of a television set, I don't recommend this project for beginners. Do not attempt a project of this nature unless you are confident you know what you are doing.

You should select the reference voltage so that the comparator (IC1) triggers reliably when it receives a signal. When there is no signal (easily simulated by tuning the TV's tuner to a channel that is not used in your area), the relay (K1) should be cut off.

You should insert the relay's contacts into the television set's power line. When there is no received signal, the relay deactivates, and the power supply is cut off from the TV set. As long as a transmitted signal is being received, the relay remains activated, allowing power to reach the set.

You might well want to include a bypass switch for this circuit, even though it is not shown in the schematic diagram.

PROJECT 19: PFM INFRARED TRANSMITTER

Traditionally, wireless communications systems usually have relied upon rf (radio-frequency) signals. Unfortunately, rf circuits usually are very sensitive to component values, parts placement, and tuning. Rf circuits also tend to have problems with interference and noise. They can interfere with other equipment and the FCC regulates them very strictly. A careless hobbyist could end up paying a hefty fine if he used too much power and/or the wrong rf frequency.

Today there is an alternative. You can transmit communications on a light beam. This approach is not perfect, of course. External light sources can interfere with the desired signal, the range is limited, and the signal path can be only line of sight. To transmit a light signal around a corner or past an opaque object of any kind, you must use a system of carefully placed mirrors or a fiberoptic cable.

Figure 4-5 illustrates the principle of a fiberoptic cable. The refractive index of the cladding (outer skin) of the cable is higher than that of the interior. This keeps the light energy confined within the interior of the fiberoptic cable. Of course, if a fiberoptic cable is used, then you no longer have a wireless communications system.

Using an infrared light source to carry the beam eliminates most interference problems. If a visible light beam is used, the communications system can be used only in a darkened area.

I discuss the receiver for this system in Project 20. For now, we will concentrate on the transmitter circuit, which is illustrated in Fig. 4-6. The parts list for this project appears in Table 4-6.

Note that the light emitter labelled D1 in this circuit is not a standard LED. It is an infrared emitter. It looks like an ordinary diode and is used in the same way, but it puts out infrared light.

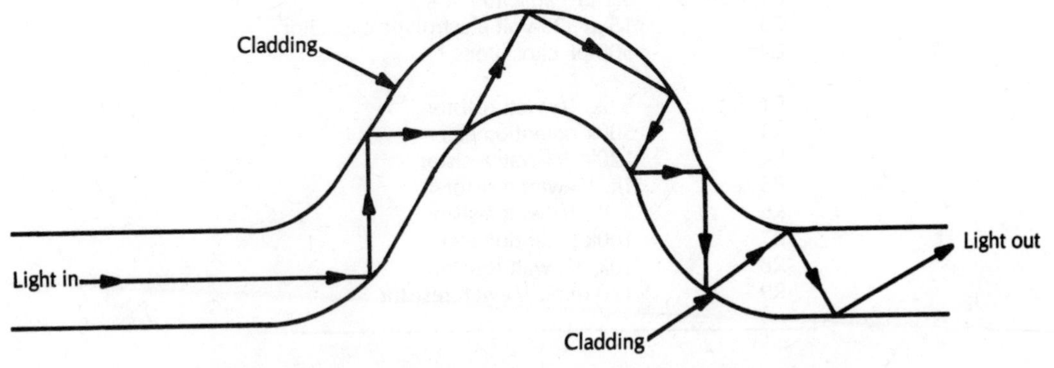

Fig. 4-5 *Encoded light beams can be carried by a fiberoptic cable.*

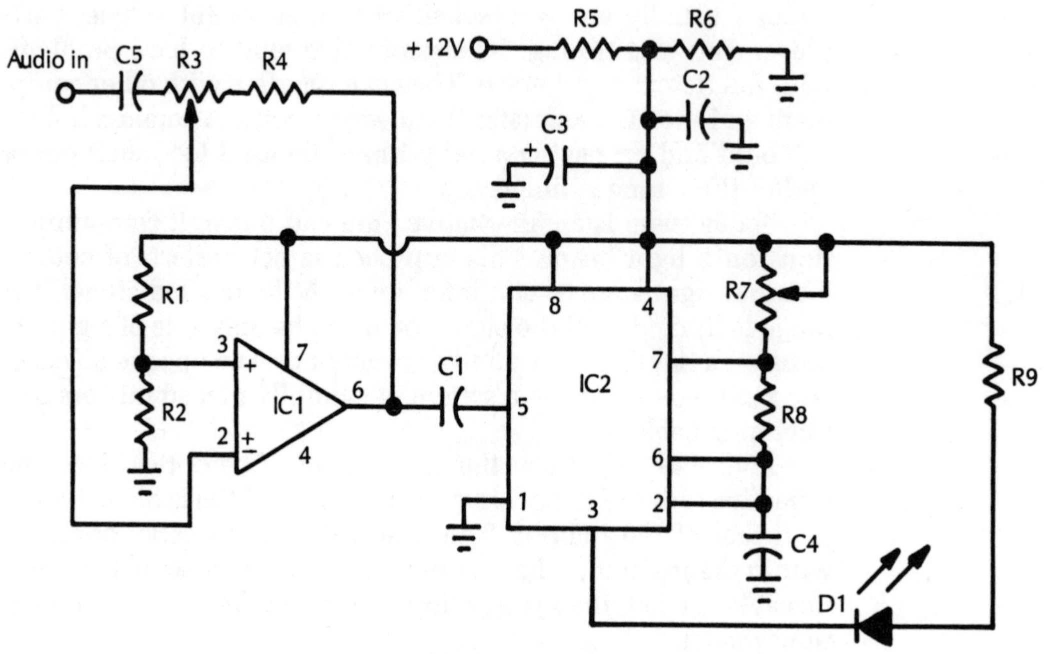

Fig. 4-6 *Project 19: PFM Infrared Transmitter.*

Table 4-6 Parts List for the PFM Infrared Transmitter of Project 19.

Part	Component
IC1	Op amp
IC2	555 timer
D1	Infrared emitter
C1, C2, C5	0.1-μF capacitor
C3	15-μF, 25-volt electrolytic capacitor
C4	500-pF capacitor
R1, R2	5.6k, 1/4-watt resistor
R3	500k potentiometer
R4	120k, 1/4-watt resistor
R5	1k, 1/4-watt resistor
R6	3.3k, 1/4-watt resistor
R7	100k potentiometer
R8	10k, 1/4-watt resistor
R9	120-ohm, 1/4-watt resistor

This transmitter circuit accepts an audio-input signal and uses it to modulate the infrared light beam. The type of modulation used in this system is known as *pulse-frequency modulation*, or PFM.

There are two calibration controls in this transmitter circuit: potentiometers R3 and R7. You can use either standard, full-sized, manually adjustable potentiometers or miniature screwdriver-adjustable trimpots for these controls, depending on your application.

Potentiometer R3 adjusts the gain of the transmitter. The higher the gain, the longer the range. However, too much gain could result in distortion in the receiver.

You should adjust potentiometer R7 for minimum distortion from the receiver. Set this control for the clearest possible sound. Actually, what this control does is adjust the carrier frequency of the transmitted signal.

The receiver for this communications system follows in Project 20. Obviously, Projects 19 and 20 are intended to be used together. One isn't much good without the other.

PROJECT 20: PFM INFRARED RECEIVER

This project is the pulse-frequency-modulated receiver to go with the transmitter described in the preceding section (Project 19) of this chapter. The schematic diagram for the PFM infrared receiver project appears in Fig. 4-7. Table 4-7 shows the parts list for this project.

IC1 can be almost any op-amp chip. For many applications, you can even use a garden-variety device, such as the 741. For more serious communications, the op amp should be a high-grade, low-noise type.

The sensor (Q1) is an npn-type phototransistor. Be sure to use a type that is sensitive to infrared rather than visible light.

Focus the infrared beam from the transmitter's emitter (infrared LED) directly onto the sensing surface of the phototransistor. A system of lenses might help in some applications. For the best results, the phototransistor (Q1) should be equipped with a shield of some sort to minimize interference from external light sources.

As I discussed in the preceding section of this chapter, if you use a fiberoptic cable to interconnect the transmitter and the

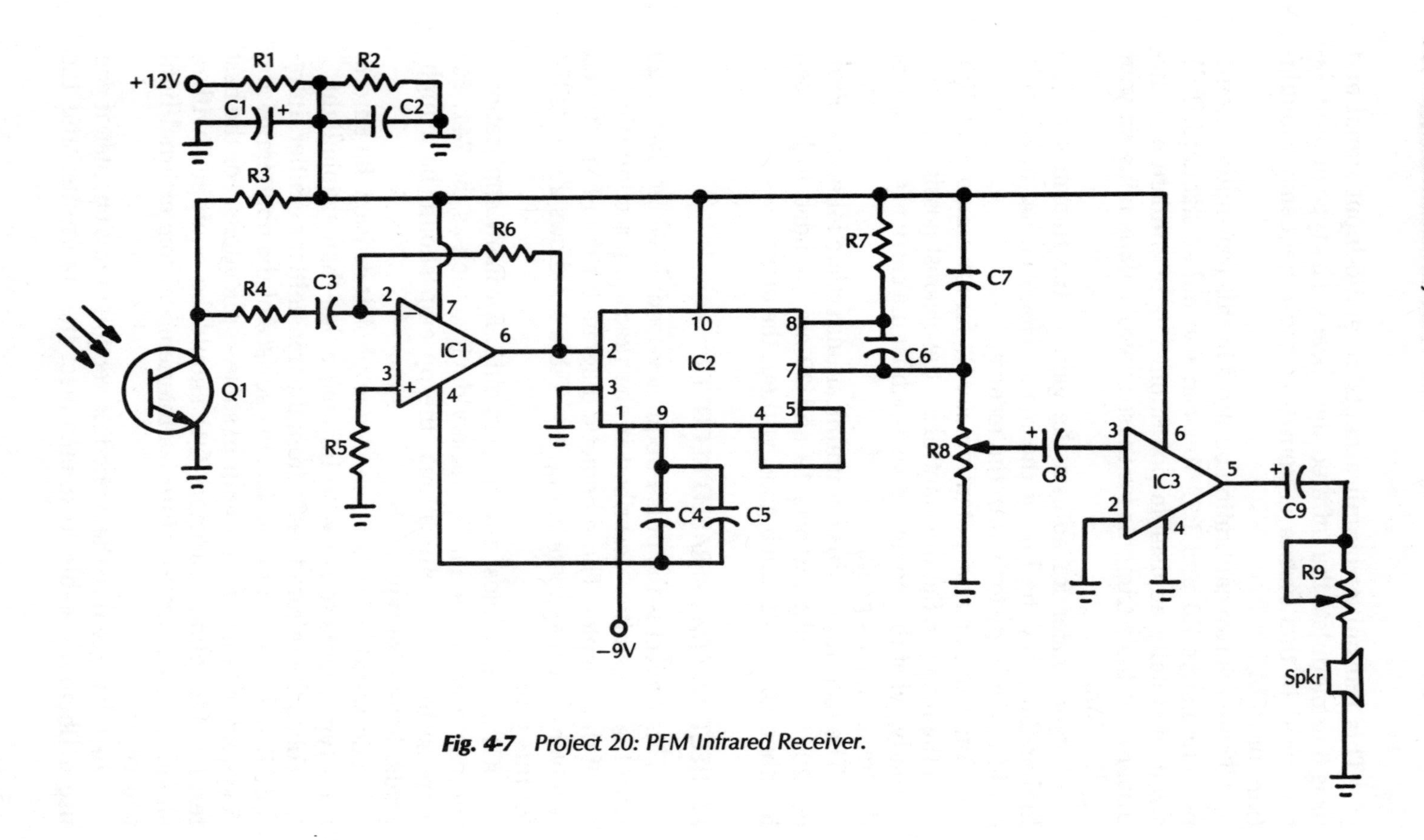

Fig. 4-7 Project 20: PFM Infrared Receiver.

Table 4-7 Parts List for the PFM Infrared Receiver of Project 20.

Part	Component
IC1	Op amp
IC2	565 PLL
IC3	386 audio amplifier
Q1	Phototransistor
C1, C2	5-μF, 20-volt electrolytic capacitor
C3	0.1-μF capacitor
C4, C5, C6	0.001-μF capacitor
C7	0.05-μF capacitor
C8	10-μF, 20-volt electrolytic capacitor
R1, R4, R5	1k, 1/4-watt resistor
R2	3.3k, 1/4-watt resistor
R3	220k, 1/4-watt resistor
R6	1-megohm, 1/4-watt resistor
R7	3.9k, 1/4-watt resistor
R8	10k potentiometer
R9	500k potentiometer
Spkr	Small 8-ohm speaker or headphone

receiver, you will not be limited to a direct line-of-sight signal path. However, by using a connecting cable, you lose the important advantages inherent in a wireless communications system.

There is no right or wrong answer here. Neither approach is inherently better or worse than the other. The choice ultimately depends on your specific application for these projects.

To align the system, feed a continuous tone source (such as an oscillator or the output from a tape recorder) into the input of the transmitter. This will make the alignment process as simple as possible. Saying "Test, test" into a microphone is too much of a nuisance. The alignment and calibration of the system would take almost forever. Be sure to start the alignment procedure with the gain on both the transmitter and the receiver at a low setting. If the signal is too weak, increase the gain by small increments.

The receiver circuit has two controls. You use potentiometer R8 to fine-tune the receiver frequency to match the transmitter. Potentiometer R9 is a simple gain, or volume, control. You can omit this control or replace it with a small fixed resistor if you choose.

The heart of the receiver circuit is in IC2. This chip is a phase-locked loop, or PLL. You can tune it with capacitors C4 and C5, and potentiometer R8 to respond to a specific carrier frequency. Once locked onto the transmitted signal, the PLL automatically readjusts itself to compensate for any minor fluctuations in the carrier frequency. This provides automatic fine tuning.

PROJECT 21: MODULATION DETECTOR

The circuit shown in Fig. 4-8 lets you know when your radio rig is tuned to an active frequency. Table 4-8 shows the parts list. This circuit recognizes a modulated signal.

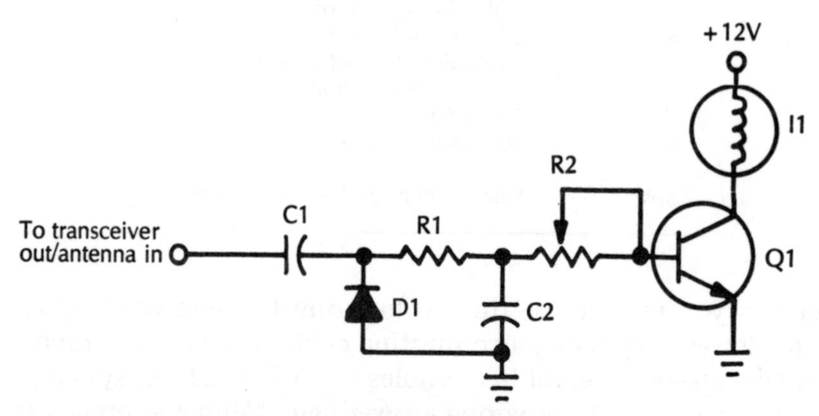

Fig. 4-8 *Project 21: Modulation Detector.*

When a modulated signal is detected, the lamp (I1) lights up. The intensity of the light put out by the lamp is proportional to the strength of the detected signal.

Adjust potentiometer R2 so that the lamp flashes in step with the modulation. For example, if the signal is a speaking voice, set R2 so that the lamp lights up whenever a word is spoken but goes dark during pauses between words or sentences. You might want to use a trimpot for R2 if you don't expect to need to reset the control frequently.

You should place this modulation monitor circuit between the transceiver and the antenna. If you are trying to transmit and the lamp (I1) doesn't light up, you know there must be something wrong with your rig.

Table 4-8 Parts List for the Modulation Detector of Project 21.

Part	Component
Q1	Npn transistor (2N3392, or similar)
D1	Germanium diode (1N60, or similar)
C1	5-pF capacitor
C2	100-pF capacitor
R1	2.2k, 1/4-watt resistor
R2	10k potentiometer
I1	12-volt, 50-mA lamp

5 ❖ Lights and Displays

CIRCUITS THAT FLASH LIGHTS ON AND OFF IN VARIOUS PATTERNS ARE always enjoyable projects. All right, so light flashers probably have somewhat limited practical applications, but why can't you build some projects just for fun? What law says electronics has to be all serious applications?

Light flashers do have some practical uses, of course. They are great eye catchers, and they can be very helpful in drawing attention to an advertisement or other type of display. They also can be used as hard-to-ignore warnings of dangerous conditions. I presented an example of this earlier in this book with the Emergency-Flasher project in chapter 2 (Project 1).

Lamp flashers also can be used to focus concentration in a biofeedback system or as an "executive relaxation device." Use your imagination, and you'll likely come up with some interesting practical application, especially if you don't want to admit that you're working on an electronics project just for kicks.

I present several intriguing light-flasher projects of various types in this chapter. Enjoy.

PROJECT 22: DUAL-LAMP FLASHER

The circuit shown in Fig. 5-1 is a very eye-catching display device. Table 5-1 shows the parts list. It alternately flashes two small lamps on and off. When lamp I1 is on, lamp I2 is off, and vice versa.

The four components in this project determine the flash rates. Both capacitors and both resistors should have identical values for equal on times for both lamps. If one of the capacitors or one of the resistors has a different value from its counterpart component, then one lamp will stay on longer than the other.

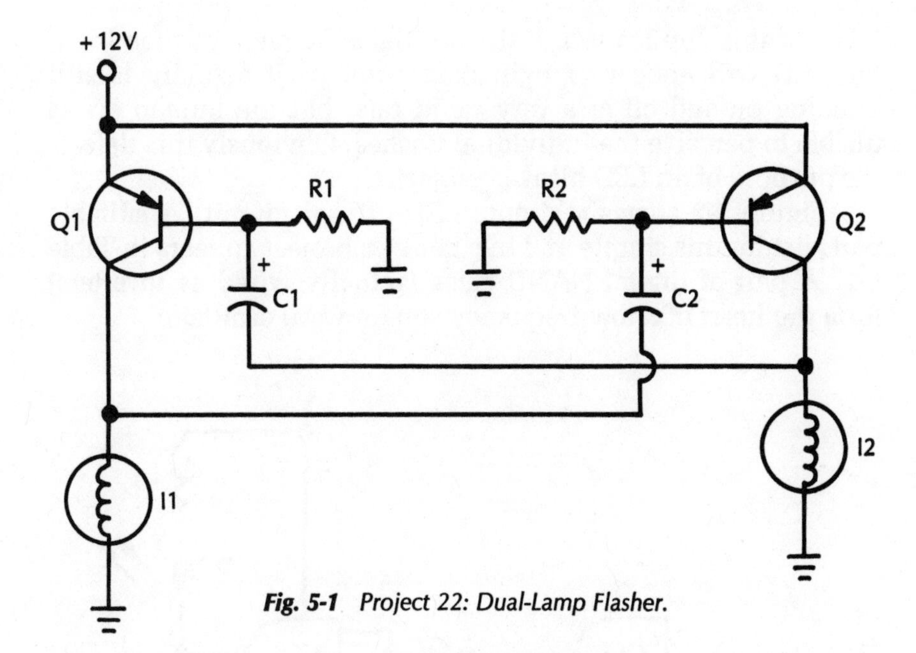

Fig. 5-1 *Project 22: Dual-Lamp Flasher.*

Table 5-1 Parts List for the Dual-Lamp Flasher of Project 22.

Part	Component
Q1, Q2	Pnp transistor
I1, I2	12-volt, 6-watt lamp
C1, C2	500-μF, 20-volt electrolytic capacitor
R1, R2	1k, 1/4-watt resistor

This might be desirable for some applications. With the component values called for in the parts list, the flash rate is about one per second. You can experiment with other values for the capacitors and resistors.

PROJECT 23: LED BLINKER

A low-speed oscillator can be used to blink an LED on and off at a regular rate. The oscillator frequency should be below 4 or

5 Hz for this application. If the oscillator frequency is too high, the LED will appear to light continuously. It actually is still blinking on and off at a very rapid rate, but the human eye is unable to perceive the individual flashes. Obviously this defeats the purpose of an LED blinker project.

Figure 5-2 shows a simple LED blinker circuit. A suitable parts list for this simple and inexpensive project appears in Table 5-2. A pair of digital NAND gates (actually wired as inverters) form the heart of a low-frequency square-wave oscillator.

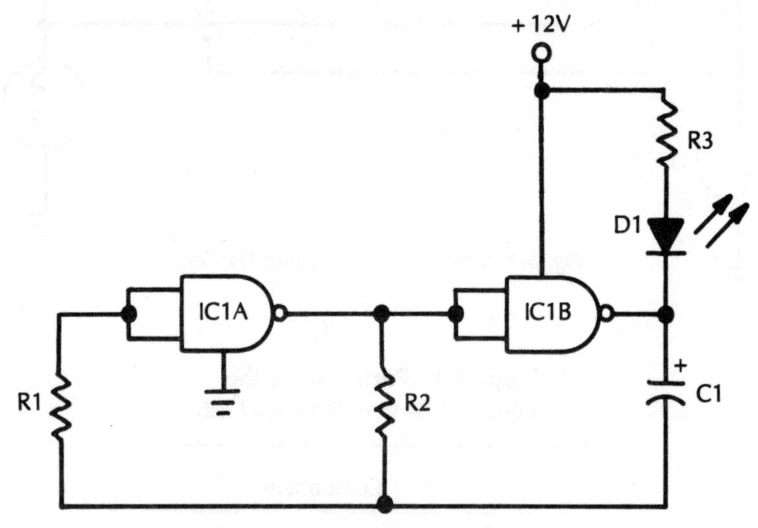

Fig. 5-2 *Project 23: LED Blinker.*

Table 5-2 Parts List for the LED Blinker of Project 23.

Part	Component
IC1	CD4011 quad NAND gate
D1	LED
C1	5-μF, 25-volt electrolytic capacitor (see text)
R1	1-megohm, 1/4-watt resistor
R2	100k, 1/4-watt resistor
R3	470-ohm 1/4-watt resistor

The component values in this project are not essential. Feel free to experiment with other component values. Changing the value of capacitor C1 has the most dramatic effect on the oscillator frequency, and thus on the flash rate. Resistor R3 limits the current flow through the LED (D1) to a safe level. With the component values in the parts list, the flash rate is about one or two blinks per second.

PROJECT 24: DUAL-LED BLINKER

Figure 5-3 shows a simple modification of the LED blinker circuit (Project 23) described above. You add another inverter stage after the oscillator to reverse its output signal. This reversed square wave is used to blink a second LED (D2) on and off.

The two LEDs blink at the same rate but in an alternate pattern. That is, when LED D1 is lit, LED D2 is dark, and vice versa. As long as you apply power to the circuit, at any given instant one of the LEDs will be on and the other one will be off.

Table 5-3 shows the parts list for this project. Note how similar this parts list is to the one for the preceding project. Because the original LED blinker circuit (Project 23) used just half of IC1, the extra inverter stage does not require the addition of another

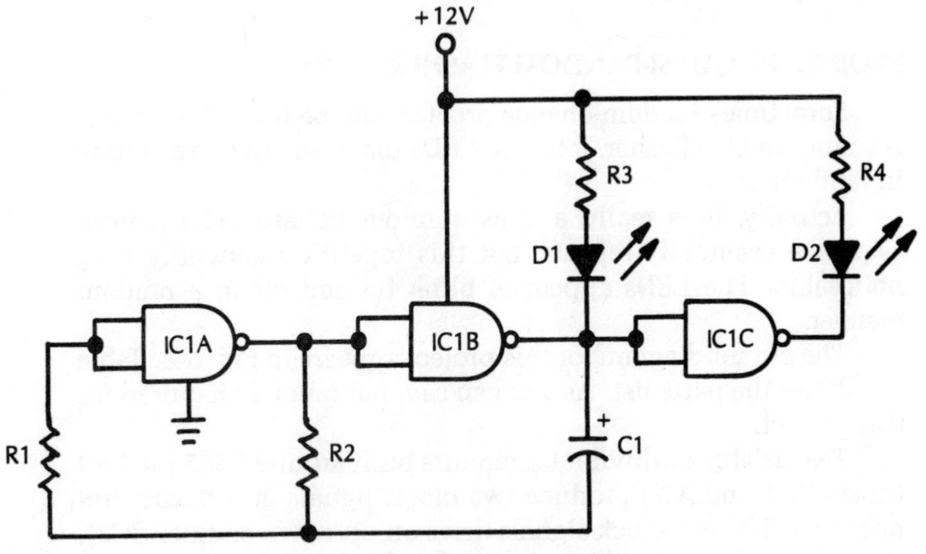

Fig. 5-3 *Project 24: Dual-LED Blinker.*

Table 5-3 Parts List for the Dual-LED Blinker of Project 24.

Part	Component
IC1	CD4011 quad NAND gate
D1, D2	LED
C1	5-μF, 25-volt electrolytic capacitor (see text)
R1	1-megohm, 1/4-watt resistor
R2	100k, 1/4-watt resistor
R3, R4	470-ohm, 1/4-watt resistor

IC. Just use one of the unused gates in IC1. The only components that need to be added for the dual LED blinker are LED D2 and resistor R4. D2, of course, is the second blinking LED. R4 is simply the current-limiting resistor for this added LED.

As with the LED blinker in Project 23, I encourage you to experiment with other component values than those given in the parts list. Especially try different values for capacitor C1. Larger capacitances slow down the flash rate, while smaller capacitance values speed it up.

PROJECT 25: QUASI-RANDOM FLASHER

Sometimes random-chance projects can be fun. This project is a random LED flasher. It has 16 LEDs that light up one at a time in random order.

Actually, it is really a quasi-random pattern. The pattern sequence eventually repeats, but this repetition generally isn't noticeable. The LEDs appear to blink on and off in a random manner.

The circuit diagram for this project appears in Fig. 5-4. Table 5-4 shows the parts list. As you can see, not much is required for this project.

Two astable multivibrator circuits built around 7555 (or 555) timers (IC1 and IC2) produce two clock signals at differing frequencies. The two clock signals drive an up/down counter (IC3). One of the clock signals increments (adds one) to the count on each pulse. The other clock signal decrements (subtracts one)

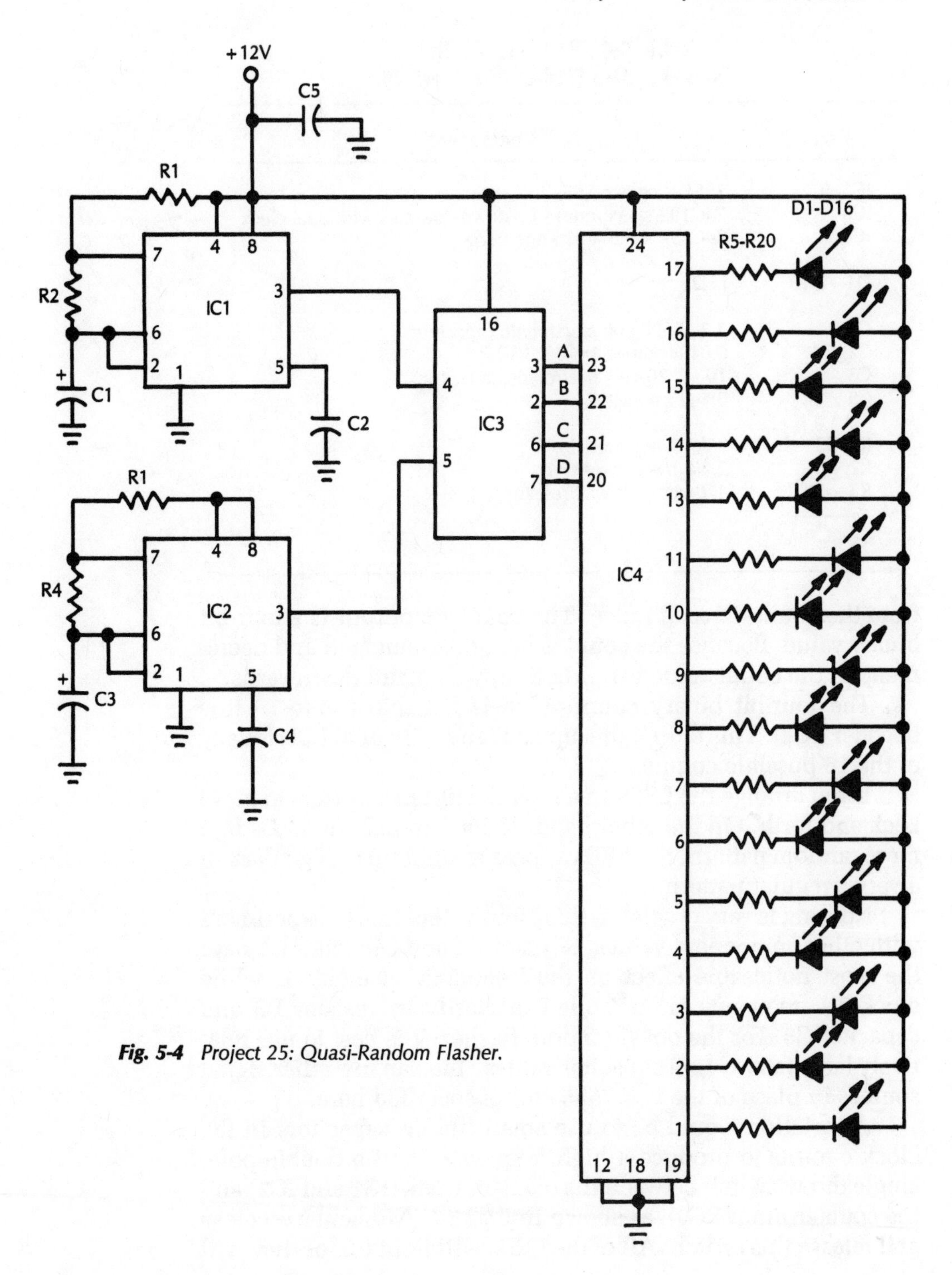

Fig. 5-4 Project 25: Quasi-Random Flasher.

Table 5-4 Parts List for the Quasi-Random Flasher of Project 25.

Part	Component
IC1, IC2	7555 timer (or 555)
IC3	74C193 synchronous up/down counters with dual clock
IC4	74C154 4-line-to-16-line decoder
D1 – D16	LED
C1	2.2-μF, 20-volt electrolytic capacitor*
C2, C4	0.01-μF capacitor
C3	10-μF, 20-volt electrolytic capacitor*
C5	0.1-μF capacitor
R1, R3	33k, 1/4-watt resistor
R2, R4	2.2k, 1/4-watt resistor
R5 – R20	470-ohm, 1/4-watt resistor

*see text

from the count on each pulse. The counter's output is a four-bit binary value. Because the count is being incremented and decremented, the count value varies both upwards and downwards.

The four-bit binary count value is fed into a 4-to-16 line decoder (IC4). This chip lights up one (and only one) LED for any of the 16 possible counts.

If you arrange the LEDs in a row, the lit LED appears to move back and forth. On the other hand, if you arrange the LEDs in a more random pattern, the LEDs appear to light up and go dark in a very irregular pattern.

Nothing is very crucial in this circuit. Feel free to experiment with other component values. Resistor R1 and capacitor C1 have the most noticeable effect on the frequency of clock 1, while clock 2's frequency is controlled primarily by resistor R3 and capacitor C3. For the quasi-random flasher, it is best to use relatively large (above 1μF) capacitor values. You can use other signal sources in place of the two 7555 clocks specified here.

A variation would be to use small timing capacitors in the clock circuits to produce a high frequency. Add a double-pole, single throw switch between the clock outputs (IC1 and IC2) and the counter input (IC3), as shown in Fig. 5-5. Momentarily close and release this switch. All of the LEDs will light up (or they will

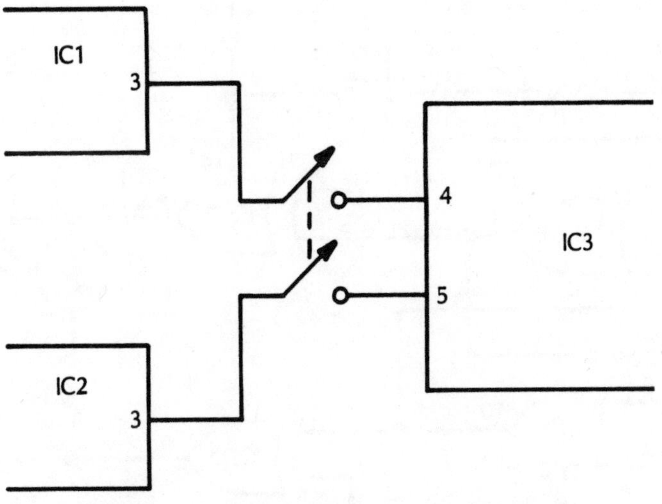

Fig. 5-5 *A suggested modification for Project 25.*

flash at a very rapid rate) as long as this switch is closed. When the switch is opened, only one of the LEDs remains lit. You can use this circuit in games, or you might want to try it as a simple "ESP tester." See if you can predict which LED will remain lit with better accuracy than pure chance would predict.

PROJECT 26: ELECTRONIC COIN FLIPPER

This project is a little bit different from the other LED flasher circuits presented in this chapter, but it is closely related. Most LED flashers blink at a fairly slow rate to make the individual flashes visible to the human eye. This project intentionally uses a very high flash rate. The LEDs flash on and off only when a controlling switch is closed.

This circuit is basically a "do-nothing box." It is definitely just for fun. You can use it as an "executive decision maker" or an electronically flipped coin for use in games.

The circuit appears in Fig. 5-6. Table 5-5 shows the parts list.

To operate this device, just momentarily close pushbutton switch S1. Both LEDs light up as long as the switch is closed. When you release the pushbutton, one of the LEDs goes dark, leaving the other one lit.

You can label one LED "Yes" and the other "No" to create an "executive decision maker." For an electronic coin flipper, you

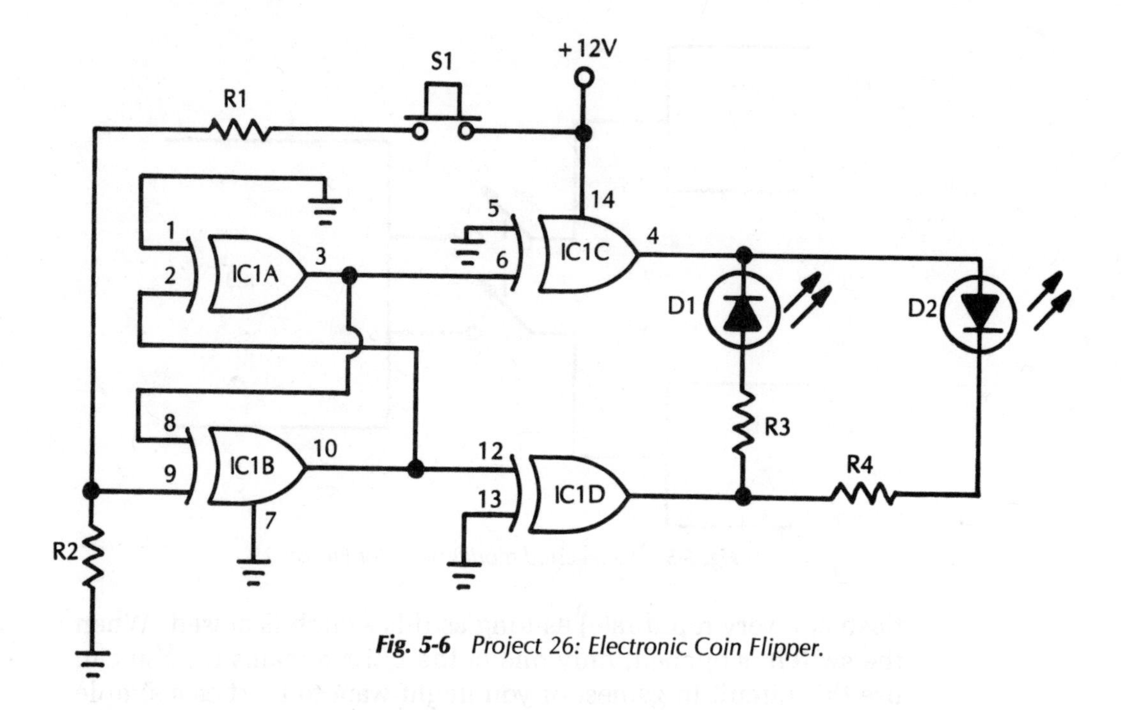

Fig. 5-6 Project 26: Electronic Coin Flipper.

Table 5-5 Parts List for the Electronic Coin Flipper of Project 26.

Part	Component
IC1	CD4077 quad X-NOR gate
D1, D2	LED
R1	1k, 1/4-watt resistor
R2	47k, 1/4-watt resistor
R3, R4	330-ohm, 1/4-watt resistor
S1	Normally open SPST push-switch

can label one LED "Heads," and the other "Tails." Use your imagination to come up with other applications and meanings for the two LEDs.

This might not be the most practical circuit in this book, but it can be a lot of fun.

❖6 Sound and Audio Projects

PROJECTS THAT GENERATE VARIOUS SOUNDS HAVE ALWAYS BEEN particular favorites of mine. This chapter focuses on signal generator and audio projects of various kinds. The circuits presented here range from amplifiers and simple oscillators or tone generators to musical instruments and accessories.

PROJECT 27: PANNING AMPLIFIER

Figure 6-1 shows an amplifier circuit that you can use to place any mono (single channel) signal source anywhere within a stereophonic field. The parts list for this project appears as Table 6-1. You can move, or pan, the source of the sound from far right to far left, and back again.

Note that IC1A and its associated components are duplicated by IC1B and its associated components. This is because this circuit is a stereo, or *two-channel*, amplifier. There is only a single input, however. A resistive voltage-divider network made up of R1, R3, and R4 splits the single-input signal into two to be fed into the separate channel amplifiers. Moving the wiper of potentiometer R3 controls how much of the input signal reaches either of the two amplifier stages. If you adjust R3 to the exact middle of its scale, each amplifier stage receives one-half of the input signal. Because their inputs are equal, their outputs are also equal. The apparent sound source appears to be located at the midpoint between the two speakers.

At one extreme of R3's rotation, virtually all of the signal is fed to amplifier stage A. No signal (or a very weak signal) is fed to amplifier stage B. The sound comes entirely from speaker A. At

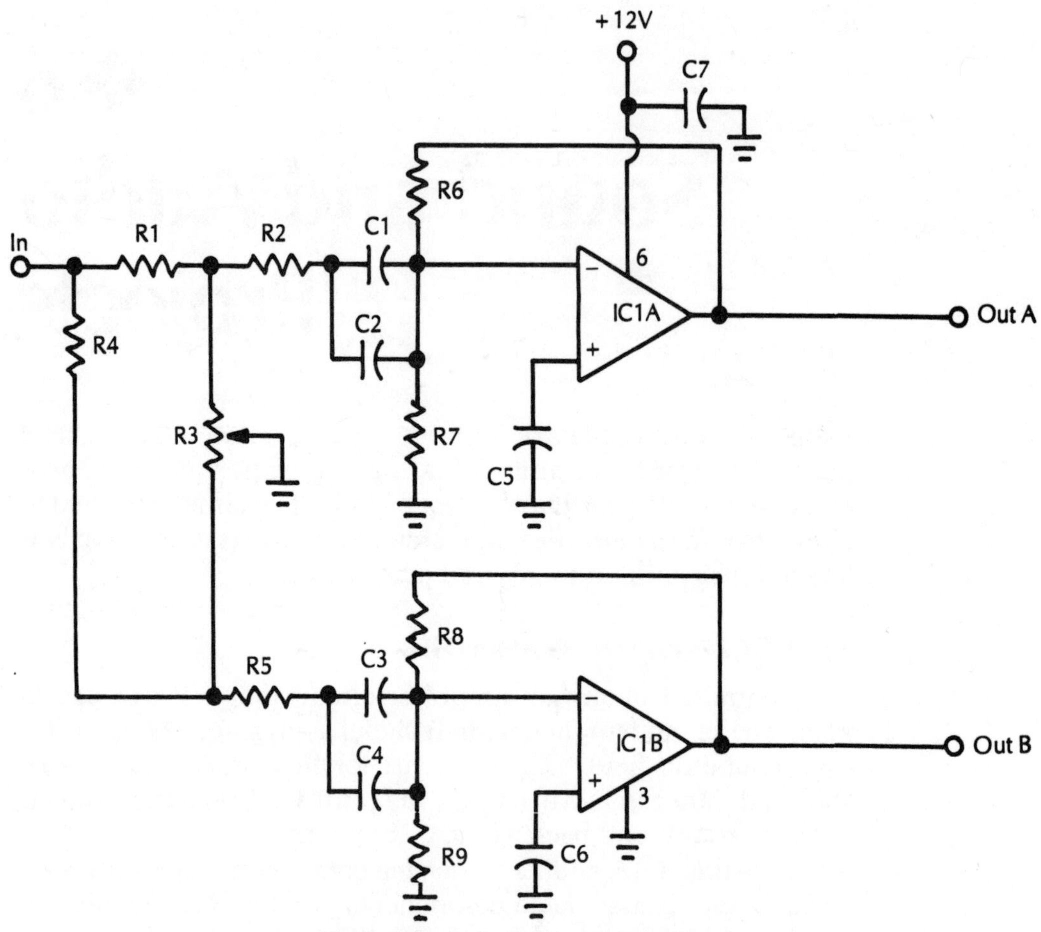

Fig. 6-1 *Project 27: Panning Amplifier.*

Table 6-1 Parts List for the Panning Amplifier of Project 27.

Part	Component
IC1	LM387 dual amplifier
C1–C4	0.47-μF capacitor
C5–C7	0.1-μF capacitor
R1, R2, R4, R5	12k, 1/4-watt resistor
R3	10k potentiometer
R6, R8	39k, 1/4-watt resistor
R7, R9	4.7k, 1/4-watt resistor

the opposite extreme of R3's rotation, you have just the opposite situation. All of the sound comes from speaker B. At intermediate points in R3's rotation, you can set the apparent sound source at any point between the two speakers. Figure 6-2 illustrates the operation of this panning-amplifier circuit.

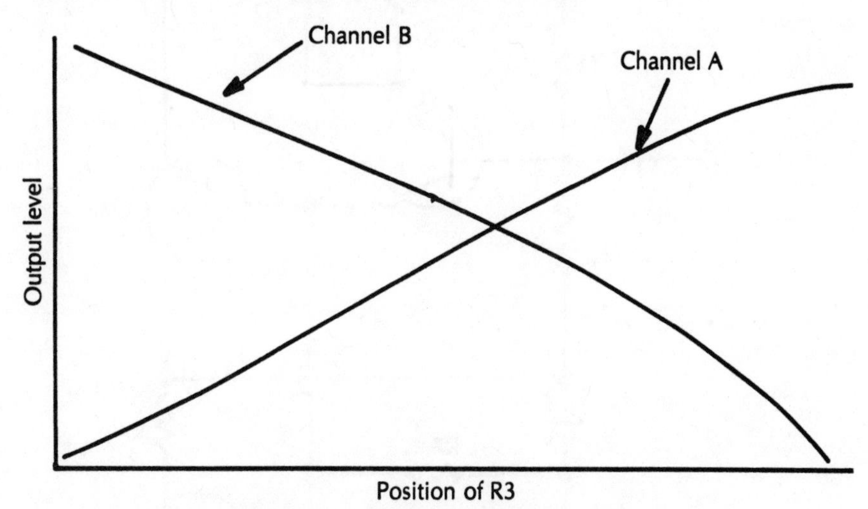

Fig. 6-2 *A panning amplifier is used to locate a sound source within a stereophonic field.*

A panning amplifier like this can be very handy in a home-recording studio. You can create a lot of interesting special effects with this circuit.

PROJECT 28: HALF-WATT HIGH-FIDELITY AMPLIFIER

There are many low-power amplifier circuits available to the electronics hobbyists. Most small amplifier circuits, however, tend to have only fair fidelity at best. High-power amplifiers are designed carefully to amplify at high fidelity and to minimize all distortion. Low-power amplifier circuits are usually "quick and dirty." They are functional, but the sound is only mediocre. In many cases, it's out and out bad.

The circuit shown in Fig. 6-3 is an exception to this. The parts list for this project appears in Table 6-2. This is a high-fidelity amplifier circuit. Because it is designed to put out just half a watt, however, it requires no expensive, heavy-duty components.

It is advisable to use the highest grade (lowest tolerance) components you can find. You might even want to use 1-watt

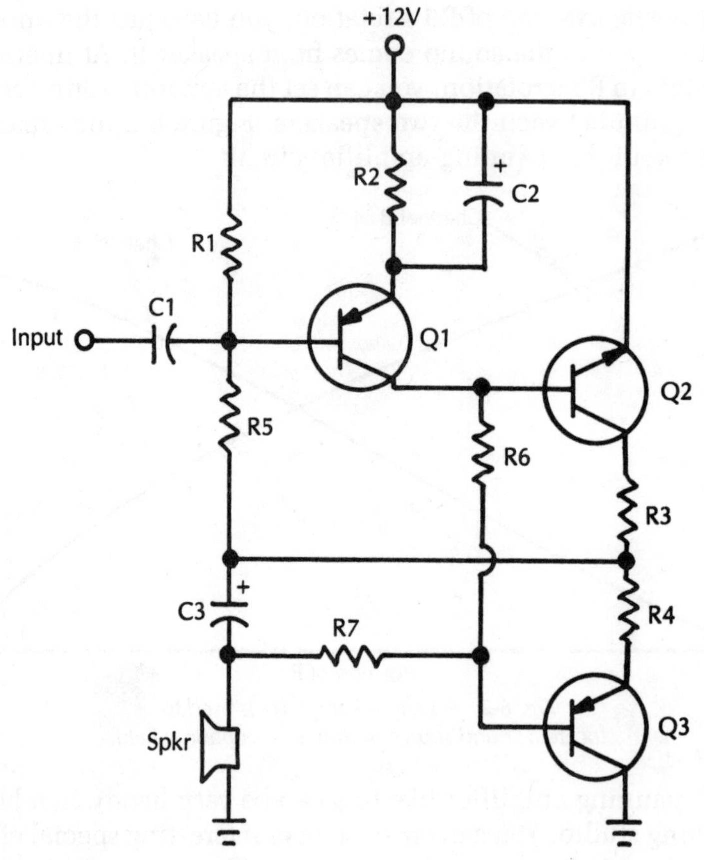

Fig. 6-3 *Project 28: Half-Watt High-Fidelity Amplifier.*

resistors instead of the 1/2-watt units in the parts list. Do not use 1/4-watt resistors in this project.

Assuming the components are close to their nominal values in the parts list, the distortion for this amplifier should be no more than 2.5 percent. The frequency response of the circuit is nominally from 20 Hz up to 100 kHz. This is well beyond the audible range. Most people can't hear frequencies above 15 kHz to 20 kHz. The frequency response is flat within ± 3.5 dB.

The speaker should be rated for an impedance of 16 ohms. Do not use an 8-ohm speaker for this circuit.

Of course, you can build two of these amplifier circuits in a single housing (with a common power supply) to create a simple low-power, but high-fidelity, stereo-amplifier system.

Table 6-2. Parts List for the Half-Watt High-Fidelity Amplifier of Project 28.

Part	Component
Q1	Pnp transistor (2N2429, or similar)
Q2	Npn transistor (2N430, or similar)
Q3	Pnp transistor (2N2706, or similar)
C1	0.5-μF capacitor
C2, C3	250-μF, 15-volt electrolytic capacitor
R1	2.2k, 1/2-watt resistor
R2	150-ohm, 1/2-watt resistor
R3, R4	10-ohm, 1/2-watt resistor
R5	12k, 1/2-watt resistor
R6	820-ohm, 1/2-watt resistor
R7	39-ohm, 1/2-watt resistor
Spkr	Small 16-ohm speaker

PROJECT 29: SMALL AUDIO-POWER AMPLIFIER

There are a number of audio-amplifier ICs available. Most tend to be fairly low-power devices. An integrated circuit cannot dissipate much heat (power) because of its small size and because its internal components are packed so tightly together. The LM383, which is shown in Fig. 6-4, can put out quite a bit of power for a device of this size. With adequate heat-sinking and an appropriate circuit, the LM383 can put out as much as 6 to 9 watts of reasonably clean audio power.

If the chip should happen to overheat, no permanent damage is likely, because the LM383 features internal thermal shutdown circuitry. It will not operate with an excessive load.

A large heat sink is essential in this project. Smear the heat sink with some heat sink compound to ensure maximum thermal transfer.

The circuit for a functional audio power amplifier is illustrated in Fig. 6-5. Table 6-3 shows the parts list.

Although two speakers are shown here, this is a monophonic (not stereo) amplifier. Note that the two speakers are wired in parallel. They are both fed the exact same signal. Two paralleled eight-ohm speakers are used here because the LM383

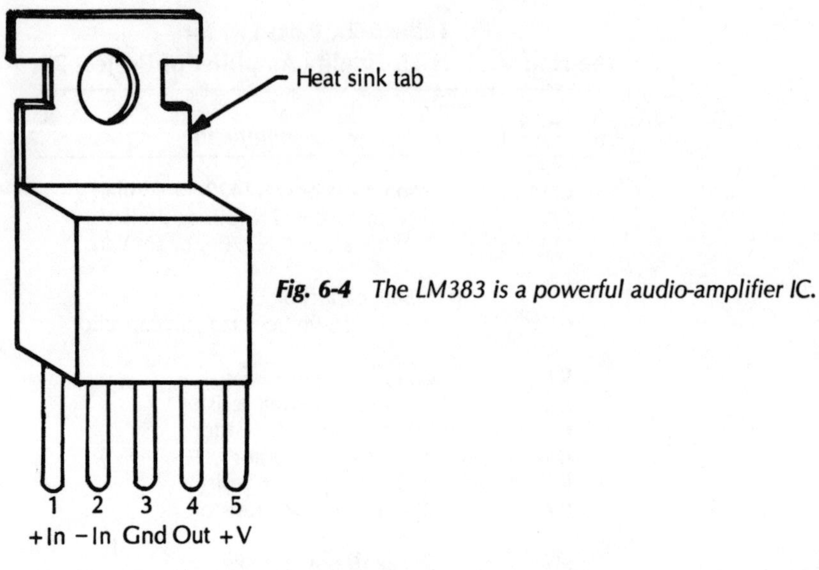

Fig. 6-4 *The LM383 is a powerful audio-amplifier IC.*

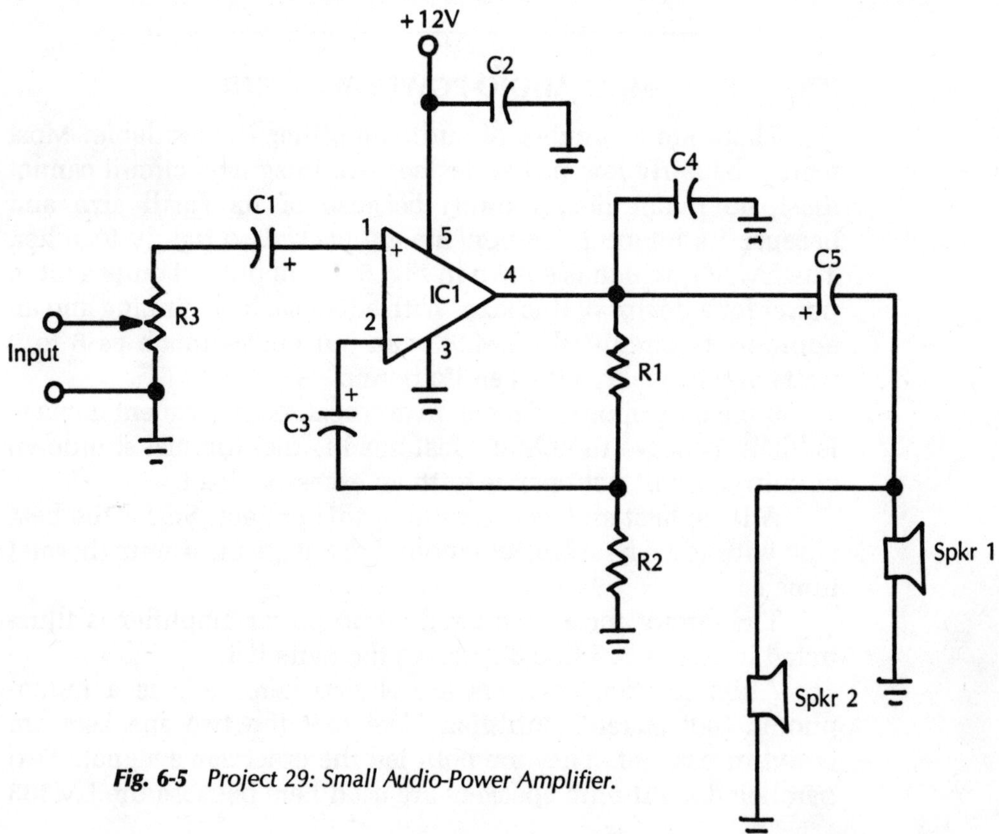

Fig. 6-5 *Project 29: Small Audio-Power Amplifier.*

Table 6-3 Parts List for the Small Audio-Power Amplifier of Project 29.

Part	Component
IC1	LM383 audio amplifier
C1	10-μF, 25-volt electrolytic capacitor
C2	0.1-μF capacitor
C3	500-μF, 25-volt electrolytic capacitor
C4	0.22-μF capacitor
C5	1000-μF, 35-volt electrolytic capacitor
R1	220-ohm, 1-watt resistor
R2	2.2-ohm, 1-watt resistor
R3	100-ohm potentiometer
Spkr 1, Spkr 2	8-ohm speaker

is designed for a four-ohm output. Eight-ohm speakers generally are easier to find than four-ohm units. If you can find a four-ohm speaker, you certainly can use it in place of the two eight-ohm speakers shown in the schematic. You should use only a single four-ohm speaker.

Alternately, you could use one eight-ohm speaker and replace the other parallel speaker with an eight- to ten-ohm, ten-watt resistor.

This circuit is quite easy to use. The potentiometer serves as a volume control.

PROJECT 30: DIGITAL CRYSTAL OSCILLATOR

A circuit for generating a continuous tone with a specific waveshape is known as a *signal generator,* or *oscillator.* Most true oscillator circuits generate a sine wave, like the one illustrated in Fig. 6-6, or a reasonable approximation of this waveform.

Ordinarily, digital circuits only use rectangle waves. The circuit shown in Fig. 6-7 uses three digital inverter stages to form a crystal oscillator with a sine-wave output. The parts list for this simple, but versatile project appears in Table 6-4.

Note that unlike many oscillator circuits, you don't need any coils (inductors) in this circuit. You can use it in place of any ordinary (linear circuit) high-frequency crystal oscillator.

The output frequency of this circuit is determined by the crystal you use.

PROJECT 31: ODD-WAVESHAPE GENERATOR

Basic oscillators and signal generators put out very simple and basic waveforms. The circuit shown in Fig. 6-8 can generate a number of very unusual waveforms and sounds.

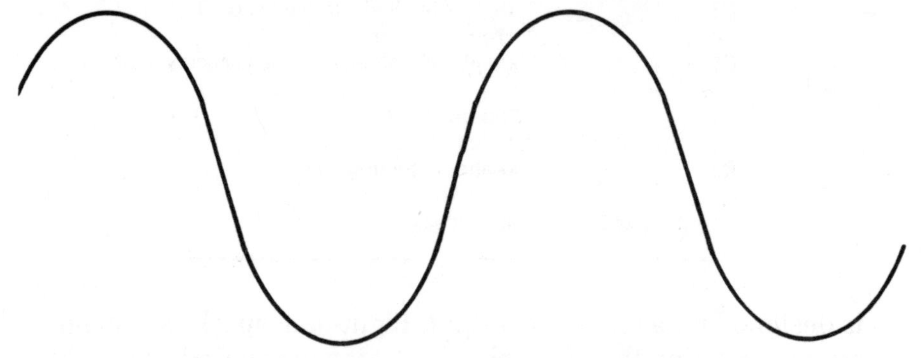

Fig. 6-6 True oscillators normally generate sine waves.

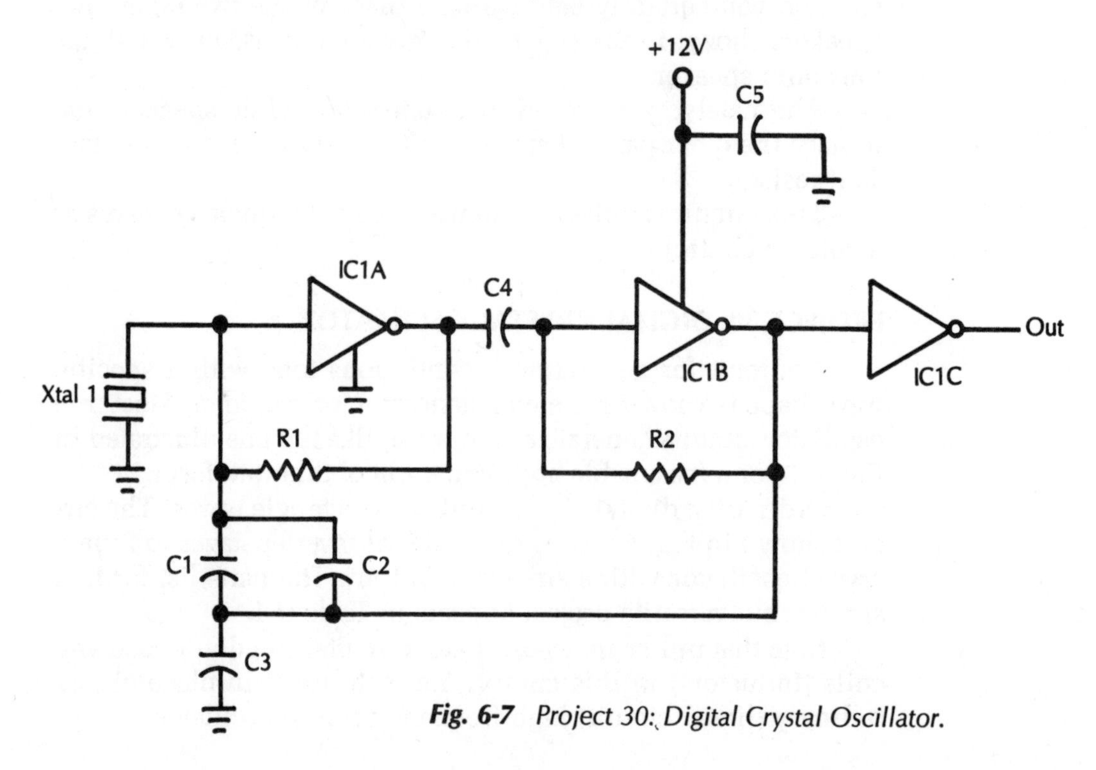

Fig. 6-7 Project 30: Digital Crystal Oscillator.

Table 6-4 Parts List for the Digital Crystal Oscillator of Project 30.

Part	Component
IC1	CD4049 hex inverter
C1	100-pF capacitor
C2	10-pF capacitor
C3	18-pF capacitor
C4	30-pF capacitor
C5	0.001-μF capacitor
R1, R2	470-ohm, 1/4-watt resistor
Xtal	Crystal (cut to desired frequency—see text)

The three timers (IC1, IC2, and IC3) generate rectangle waves of various frequencies. The circuit functions best with 7555 CMOS timers, but you can use standard 555 timers without problems. No circuit changes are required to make this substitution.

The outputs from the three rectangle-wave generators are combined by the four gates of IC4. The output is a complex, irregular pulse pattern. A number of very unusual sounds can result, depending on the specific frequency and the phase relationships of the three rectangle waves. Often there is no definite, perceptible pitch in the output signal.

For an even wider range of possible sounds, feed the output through one or more filters or other signal-altering devices (such as guitar effects boxes) before feeding the signal into an amplifier for audio production.

You can manually control the frequencies of the three input rectangle waves with potentiometers R1, R3, and R5.

A typical parts list for this project appears in Table 6-5. Nothing is particularly essential in this project. Feel free to experiment with other component values, especially for the frequency-determining capacitors for each rectangle-wave generator (C1, C3, and C5). Personally, I think you obtain the best results if you set one of the rectangle-wave generators for a very low (subaudible) frequency. You do this by using a relatively large timing capacitor. For this reason, I specify capacitor C1 in the parts list at 10μF.

There is plenty of room for experimentation in this circuit.

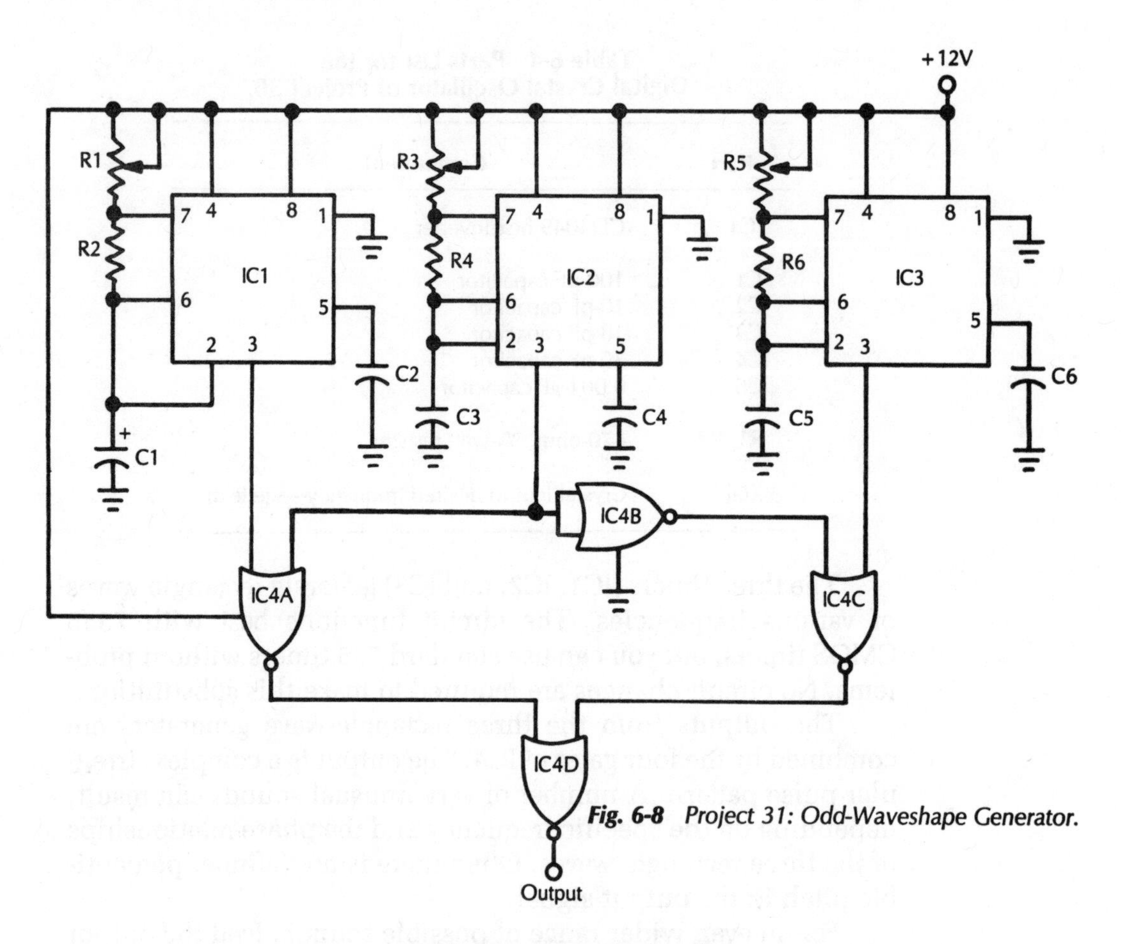

Fig. 6-8 Project 31: Odd-Waveshape Generator.

Table 6-5 Parts List for the Odd-Waveform Generator of Project 31.

Part	Component
IC1 – IC3	7555 timer (or 555)
IC4	CD4011 quad NOR gate
C1	10-μF, 15-volt electrolytic capacitor*
C2, C4, C6	0.01-μF capacitor
C3	0.1-μF capacitor*
C5	0.05-μF capacitor*
R1, R3, R5	10k potentiometer
R2, R4, R6	1k, 1/4-watt resistor

*Frequency-determining component—see text

For example, try feeding an external control voltage signal into pin 5 of one of the timers (eliminate the stabilizing capacitor—C2, C4, or C6—for that chip). This gives you dynamic control over the sound, and you can create a great many new and exciting effects.

PROJECT 32: NOISE GENERATOR

Ordinarily, we think of noise as something undesirable and to be avoided as much as possible. In some electronics applications, however, you might want to apply some noise deliberately.

In electronics, the word noise has a slightly different meaning than in ordinary, day-to-day usage. Usually when people talk about noise, they are referring to an unpleasant sound. In electronics, *noise* is a broadband signal, comprised of all frequencies (within a specific range) randomly mixed together. At any given instant, the signal is equally likely to be at any of its component frequencies.

A noise generator often is used in electronic testing. For example, an amplifier with poor frequency response will change very noticeably the audible character of the noise signal.

Noise generators also are used frequently in electronic music to create percussive, nonpitched sounds such as drums, thunder, wind, and breath.

Figure 6-9 shows a simple broadband noise-generator circuit. You can use this circuit with either audio or rf (radio-frequency) circuits. Note that the zener diode (D1) is reverse biased to force it into an avalanche condition. This is the actual noise source. The three transistors amplify the noise signal to a usable level.

The parts list for this project appears in Table 6-6. The component values are not particularly essential. If a component is widely out of tolerance, the most probable result will be a change in the balance of the various component frequencies. Certain frequencies can be more heavily represented than others.

PROJECT 33: PHASE SHIFTER

Phase shifting is a popular effect for electronic musical instruments. *Phase* is a measurement of when each cycle begins and ends, with reference to a specific standard. If two signals with the same frequency begin and end each cycle at precisely

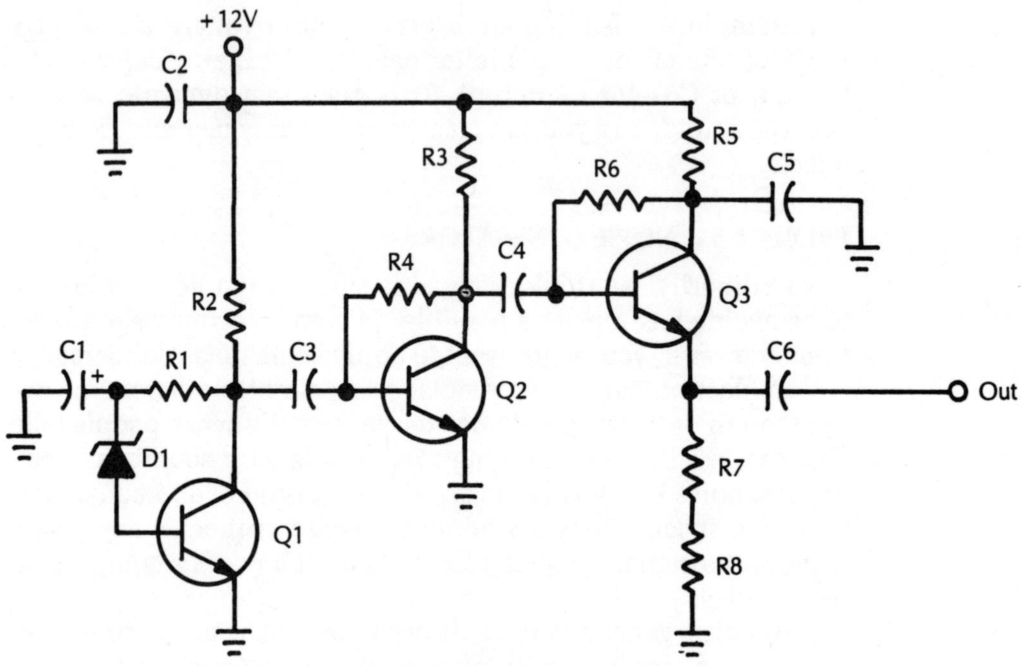

Fig. 6-9 Project 32: Noise Generator.

Table 6-6 Parts List for the Noise Generator of Project 32.

Part	Component
Q1–Q3	Npn transistor (HEP736, ECG107, SK3020, or similar)
D1	Zener diode (1N759, or similar)
C1	3.3-μF, 15-volt electrolytic capacitor
C2–C5	0.1-μF capacitor
C6	0.001-μF capacitor
R1	68k, 1/4-watt resistor
R2	6.8k, 1/4-watt resistor
R3	10k, 1/4-watt resistor
R4	390k, 1/4-watt resistor
R5	8.2k, 1/4-watt resistor
R6	330k, 1/4-watt resistor
R7	470-ohm, 1/4-watt resistor
R8	100-ohm, 1/4-watt resistor

the same time, as illustrated in Fig. 6-10, they are said to be *in phase*. The phase angle is 0°.

If one signal's cycle starts at the exact middle of the cycle of the reference signal, as shown in Fig. 6-11, they are 180° out of phase. If you mix these two signals (assuming the same waveshape and amplitude) together, the net result is zero. The two out-of-phase signals cancel each other out, leaving no signal at all.

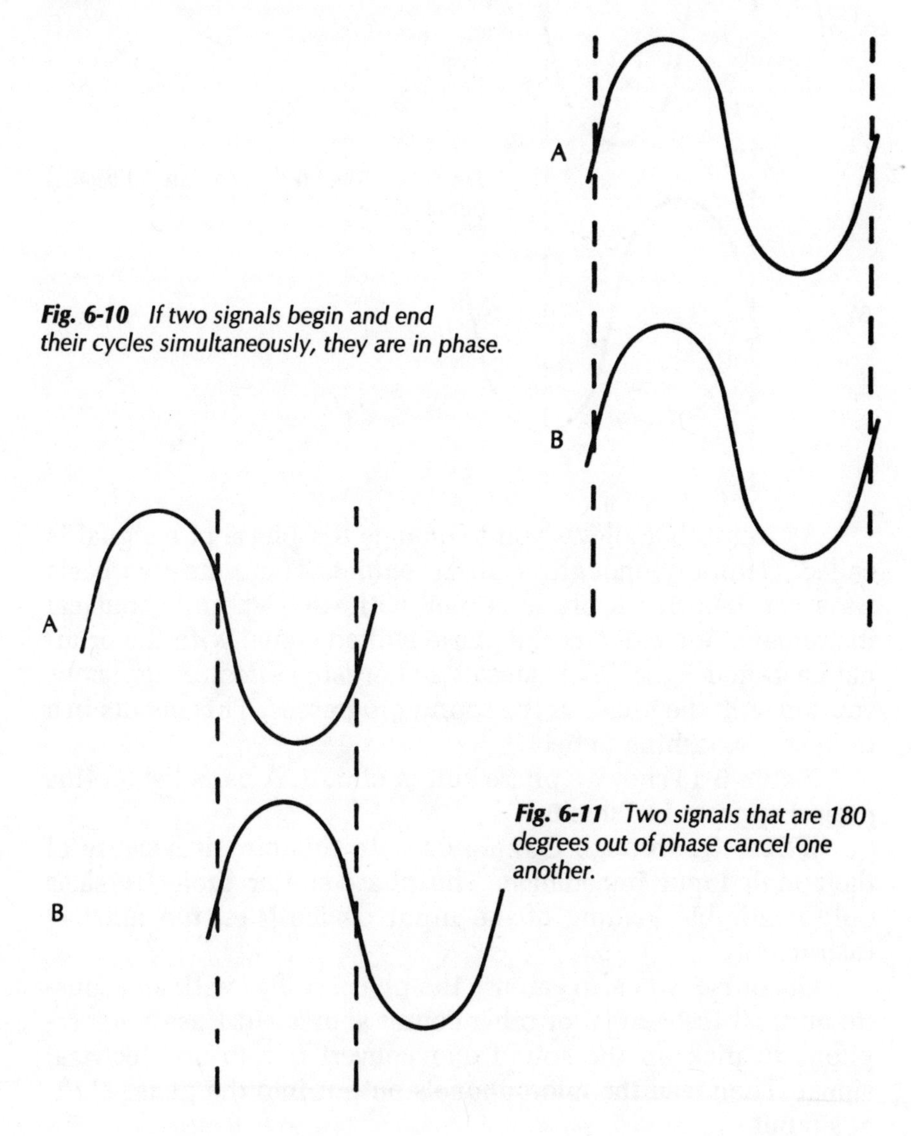

Fig. 6-10 *If two signals begin and end their cycles simultaneously, they are in phase.*

Fig. 6-11 *Two signals that are 180 degrees out of phase cancel one another.*

A full cycle is 360°. If two signals are 360° out of phase with one another, this is the same as a phase difference of 0°. In other words, the two signals are actually in phase. Two signals 90° out of phase with one another are illustrated in Fig. 6-12.

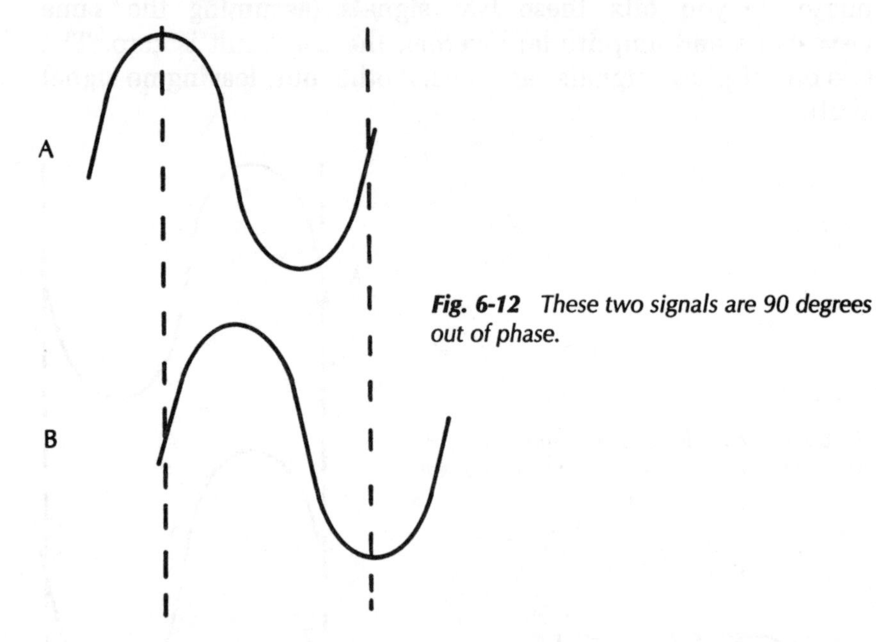

Fig. 6-12 *These two signals are 90 degrees out of phase.*

A circuit that allows you to change the phase of a signal is called, naturally enough, a *phase shifter.* There are two basic ways you can use a phase shifter with an electronic musical instrument. You can mix the phase-shifted signal with the original unshifted signal. This creates a chorusing effect. Alternately, you can shift the phase as the sound progresses. This results in a unique "swooshing" effect.

Figure 6-13 shows a phase-shifter circuit. A parts list for this project appears in Table 6-7.

JFETs (Q1 through Q3) are used in this circuit because of their high input impedance. The phase-shifter project creates only negligible loading of the input device (i.e., the musical instrument).

Of course, you also can use the phase shifter with an acoustic musical instrument or other sound source. Just use a microphone to pick up the sound and convert it into an electrical signal. Then feed the microphone's output into the phase-shifter's input.

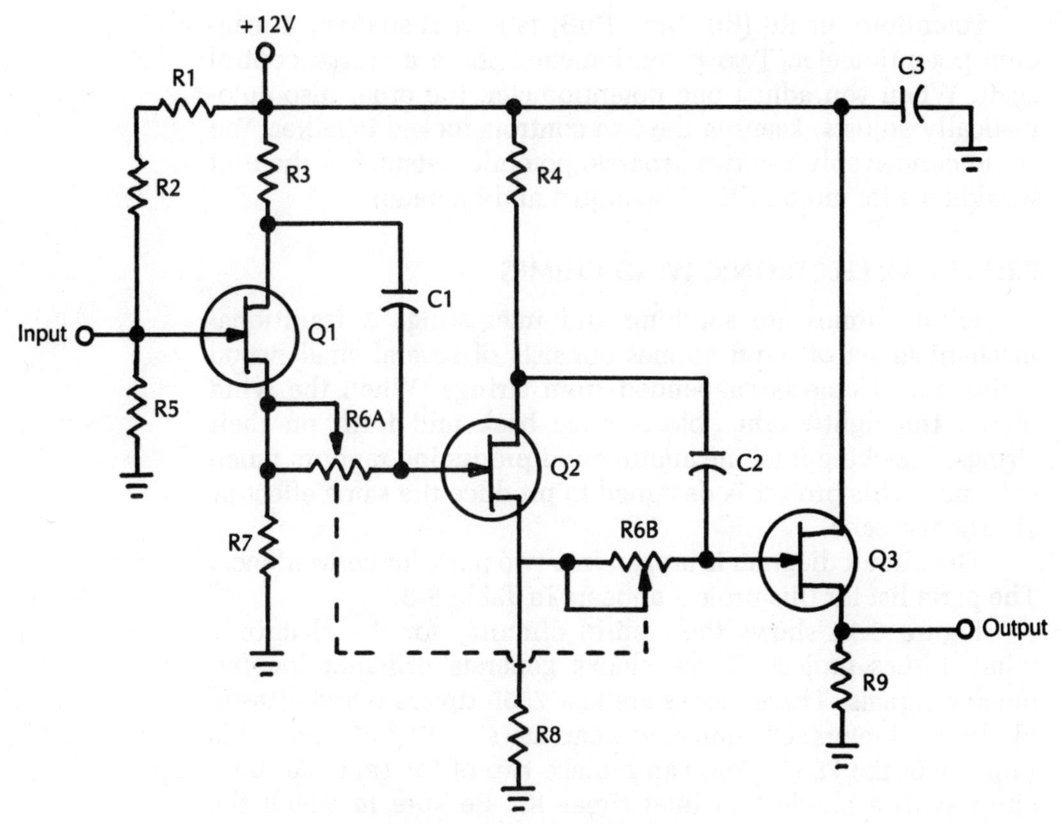

Fig. 6-13 Project 33: Phase Shifter.

Table 6-7 Parts List for the Phase Shifter of Project 33.

Part	Component
Q1, Q2	UJT (2N3070, or similar)
C1, C2	0.01-μF capacitor
C3	0.1-μF capacitor
R1, R2	3.3-megohm, 1/4-watt resistor
R3, R7	2.2k, 1/4-watt resistor
R4, R8, R9	4.7k, 1/4-watt resistor
R5	1-megohm, 1/4-watt resistor
R6	Dual 1-megohm potentiometer (see text)

Potentiometer R6 (R6A and R6B) is a dual-shafted, 1-meg-ohm potentiometer. Two potentiometers share a single control shaft. When you adjust one potentiometer, the other also automatically adjusts, keeping the two controls locked together. You could conceivably use two separate potentiometers, but the unit would be a lot more difficult to adjust and operate.

PROJECT 34: ELECTRONIC WIND CHIMES

Wind chimes are soothing and interesting. A traditional mechanical set of wind chimes consists of several small wood and/or metal objects suspended from strings. When the wind blows, the lightweight objects sway back and forth on their strings, knocking into one another and producing random musical tones. This project is designed to produce the same effect by electrical means.

The circuit diagram is broken into two parts for convenience. The parts list for this project appears in Table 6-8.

Figure 6-14 shows the control circuitry for the electronic wind-chimes project. Three clocks generate different low-frequency signals. These clocks are just 7555 timers wired as astable multivibrators. Of course, you can use standard 555 timer ICs in place of the 7555s. You can replace two of the separate timer chips with a single 556 dual timer IC. Be sure to watch the changes in pin numbering.

The frequencies of the three clocks are controlled by the appropriate potentiometer:

Timer 1	R2
Timer 2	R5
Timer 3	R8

You can replace these manual controls with trimpots or even fixed resistors if you prefer.

The timing capacitor for each clock sets the frequency range:

Timer 1	C1	10μF
Timer 2	C3	100μF
Timer 3	C5	50μF

You can experiment with other capacitor values if you like. To simulate wind-chime operation, you should use low frequencies

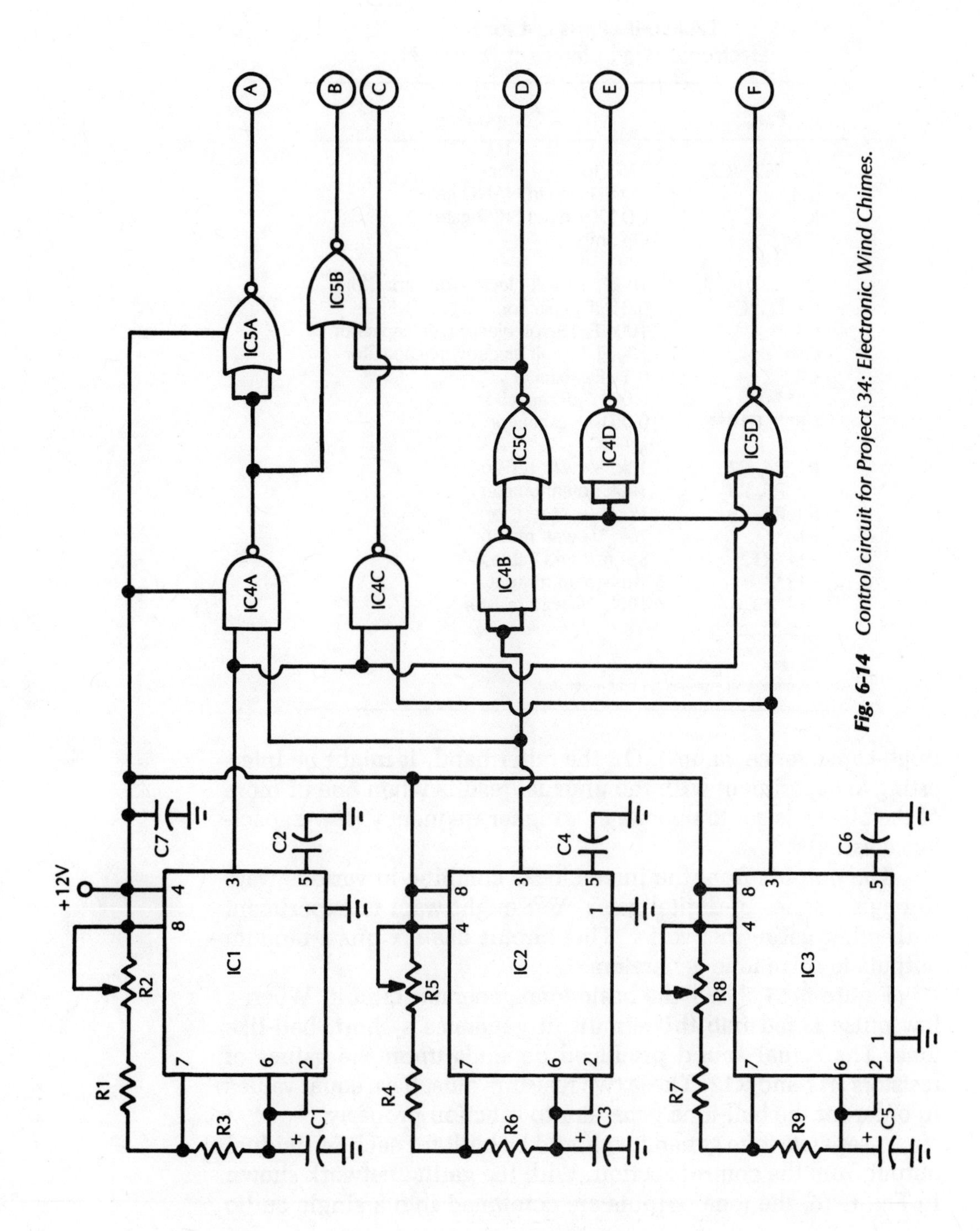

Fig. 6-14 *Control circuit for Project 34: Electronic Wind Chimes.*

Table 6-8 Parts List for the Electronic Wind Chimes of Project 34.

Part	Component
IC1, IC2, IC3	7555 (or 555) timer
IC4	CD4011 quad NAND gate*
IC5	CD4001 quad NOR gate*
IC6**	Op amp*
C1	10-μF, 15-volt electrolytic capacitor
C2, C4, C6	0.01-μF capacitor
C3	100-μF, 15-volt electrolytic capacitor
C5	50-μF, 15-volt electrolytic capacitor
C7	0.1-μF capacitor
C8**	0.0047-μF capacitor
C9**, C10**	0.001-μF capacitor
R1, R4, R7	10k, 1/4-watt resistor
R2, R5, R8	100k potentiometer
R3, R6, R9	1k, 1/4-watt resistor
R10**	10k, 1/4-watt resistor
R11, R12	See text and Table 6-9
R13**	1-megohm trimpot
R14**	100k, 1/4-watt resistor

*See text
**Component repeated six times—see text

(high-capacitance values). On the other hand, it might be interesting to experiment with the unusual results when one or more of the clocks is set to operate at a higher frequency (low-capacitance value).

The outputs from the three clocks combine in various ways through a series of digital gates. You might want to experiment with other gating networks. This circuit has six quasi-random outputs to drive tone generators.

Figure 6-15 shows the basic tone-generator circuit. When a low pulse is fed into this circuit, it generates a short, bell-like tone. The actual sound produced depends upon the values of resistors R11 and R12. These two resistors must have equal values in order for the bell-tone generator to function properly.

A separate tone generator should be built for each individual output from the control section. With the gating network shown in Fig. 6-16, the tone outputs are combined into a single audio

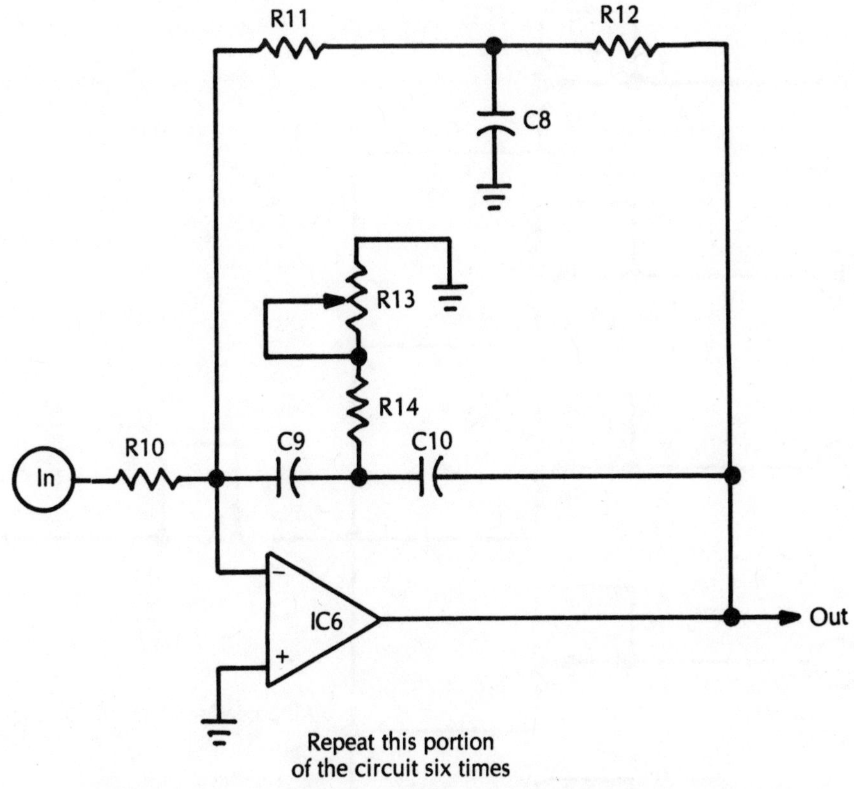

Fig. 6-15 *Tone generator for the Electronic Wind Chimes project.*

output to be fed to an external amplifier, such as Project 28 or Project 29 (both presented earlier in this chapter). You probably will want to use a different value for each individual bell-tone generator, so your electronic wind chimes can make a variety of sounds. If all the bell-tone generators sound alike, the effect will be quite monotonous and boring. The resistance value used determines the apparent size of the wind chime-object or bell. Smaller resistances result in a more tinkling effect, like a small bell or other ringing object. Larger resistance values produce more of a gonglike effect at a lower frequency.

Remember that for any given bell-tone generator, resistors R11 and R12 must have identical resistances. Table 6-9 shows some suggested resistance values for the six tone generators.

Each bell-tone generator must be calibrated individually by trimpot R13. Adjust the trimpot so that the generator breaks into oscillation (produces a continuous tone). Slowly and carefully

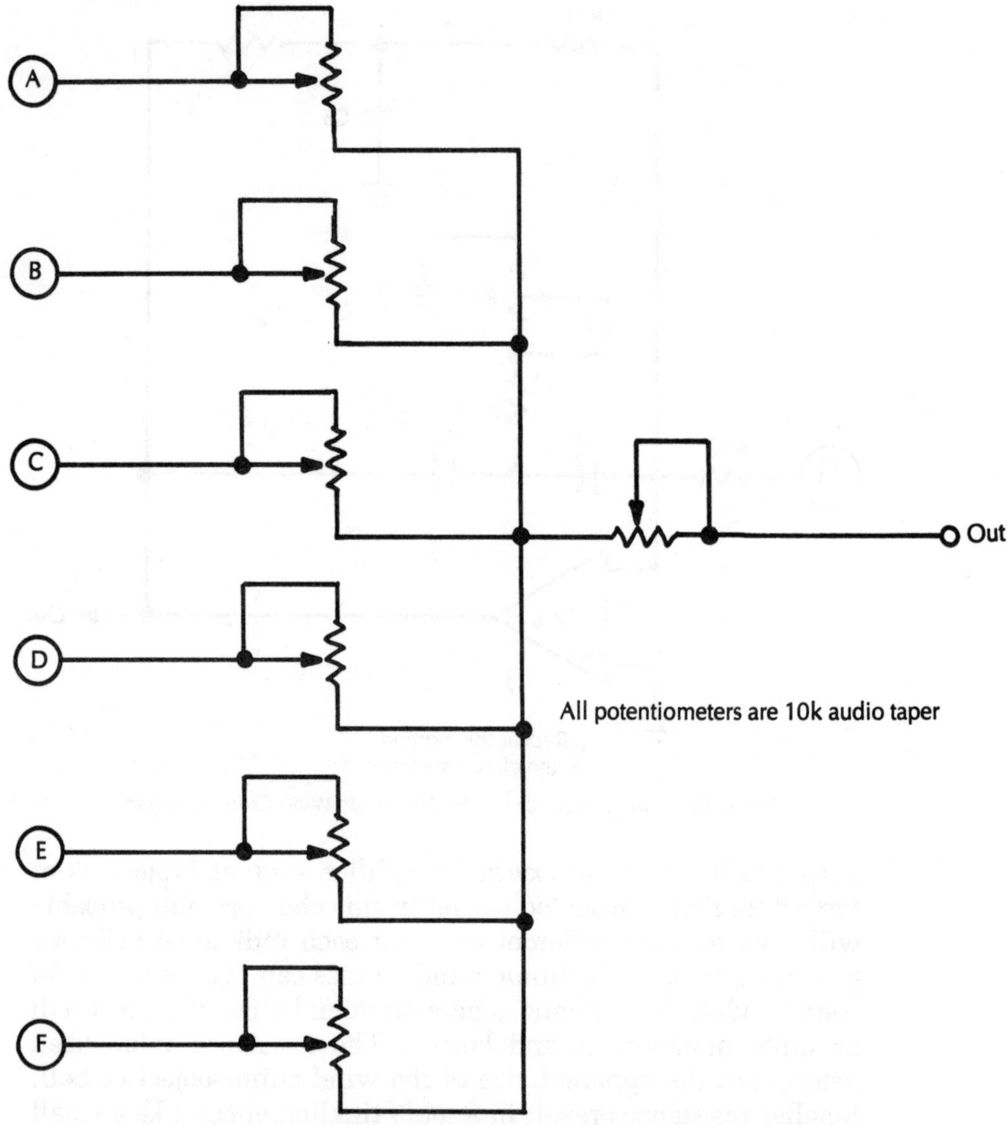

Fig. 6-16 Typical gating network for the Electronic Wind Chimes project.

back off on the adjustment of the trimpot so the oscillations just stop. Every time the generator circuit receives a pulse it produces a brief, ringing tone. This adjustment is crucial.

You can use almost any op-amp IC, including the popular 741, as the heart of the bell-tone generators (IC6). Note that no power-supply connections are shown to the op-amp. These connections always are implied, even though they might be omitted

Table 6-9 Suggested Resistor Values for the Tone Generators.

Tone	R11 and R12
A	1-megohm, 1/4-watt resistor
B	820k, 1/4-watt resistor
C	680k, 1/4-watt resistor
D	470k, 1/4-watt resistor
E	330k, 1/4-watt resistor
F	220k, 1/4-watt resistor

from schematic diagrams for reasons of greater clarity. Obviously, you must supply power to all chips in any circuit. Many op-amp ICs (including the 741) require a dual-polarity power supply. To avoid the need for a negative supply voltage, you might want to use a chip like the LM324 quad op-amp. Besides requiring just a single supply voltage, the 324 allows you to build four independent bell-tone generators around just a single IC package.

There are a lot of possible variations you can work up around this project, so feel free to experiment. I leave it to your own imagination.

7 ❖ Test Equipment

ANYBODY WHO WORKS WITH ELECTRONICS IN ANY CAPACITY, whether professionally or as an amateur, has a need for test equipment. A well-stocked electronics workbench should at least have a multimeter. This is the most useful and versatile piece of test equipment you can own. The second most frequently used test instrument is the oscilloscope. You're probably better off buying a multimeter or an oscilloscope already assembled or in kit form rather than trying to design and build your own. However, some specialized test equipment is worth building yourself.

Specialized test equipment tends to be expensive. Some types aren't even available commercially to the hobbyist. If you aren't going to use a piece of test equipment often, it doesn't make sense to buy an expensive piece of test equipment. It might be handy, however, to have a "quick and dirty" specialized test device around for those rare times when the need arises.

This chapter features a number of useful pieces of test equipment for you to build. These projects can save you a lot of hassle and frustration when it comes time to troubleshoot your other projects or electronic equipment.

PROJECT 35: CONTINUITY TESTER

In electronics work, you often need to check for continuity (an electrical path) between two points. For example, you might need to check for a break within a wire. Alternately, you might need to determine there is no short circuit between two leads.

You can use an ohmmeter for this purpose, but this would be overkill. You really aren't interested in the actual resistance value, just "is it high or low?" Besides, with a standard ohmmeter (on a VOM or DMM), you have to watch the meter or digital display. This might be inconvenient in some cases. For example, you might need to jiggle some wires during the test to find intermittent shorts or breaks.

The circuit shown in Fig. 7-1 is a handy and inexpensive solution to such problems. It is a dedicated continuity tester with an audible output. The parts list for this project appears in Table 7-1.

If there is no electrical path (continuity) between the two test leads, the output speaker remains silent. The oscillator circuit built around the 555 timer (IC1) is incomplete in this case, so it cannot function.

However, suppose there is a dead short (zero resistance) between the test leads. In this case, the circuit is virtually identical to the one shown in Fig. 7-2. This is simply a standard 555 rectangle-wave generator. The speaker emits an audible tone, and LED D1 also lights up to provide a visual indication.

This circuit also can give a rough measurement of resistance. Suppose there is a finite resistance (greater than zero) between the two test leads. There is continuity but not a dead short.

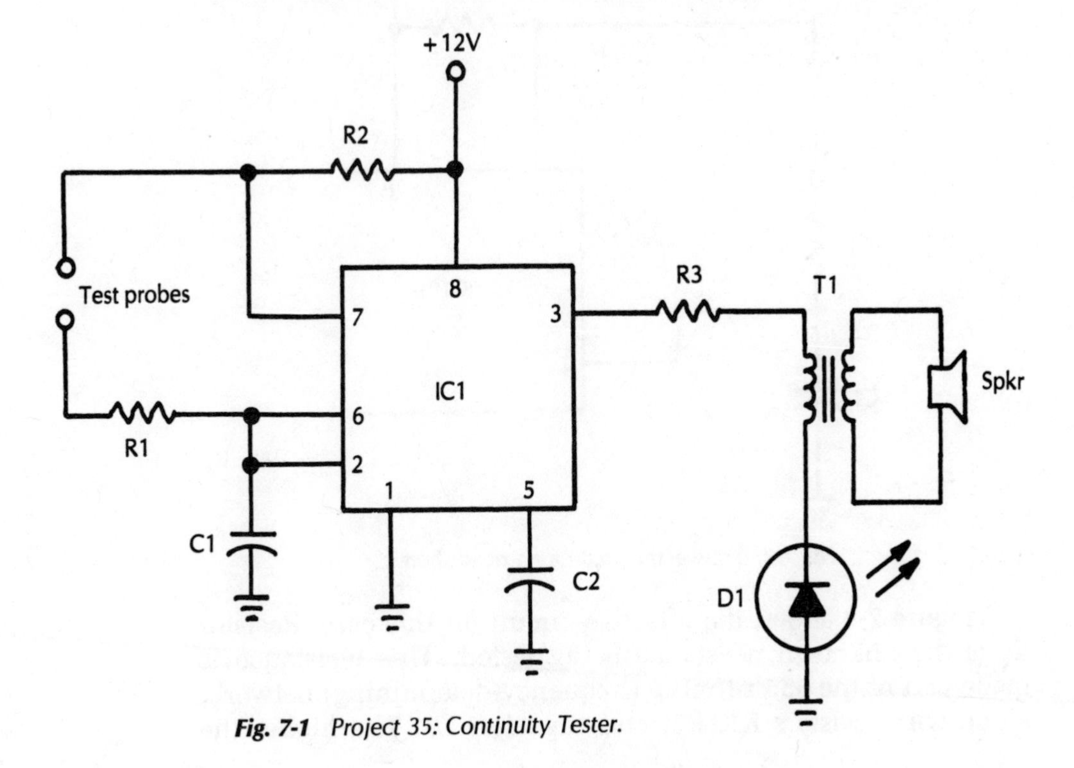

Fig. 7-1 Project 35: Continuity Tester.

Table 7-1 Parts List for the Continuity Tester of Project 35.

Part	Component
IC1	555 timer
D1	LED
C1, C2	0.01-μF capacitor
R1	10k, 1/2-watt resistor
R2	22k, 1/2-watt resistor
R3	1k, 1/2-watt resistor
T1	Audio transformer primary impedance: 500 ohms secondary impedance: 8 ohms
Spkr	Small speaker, 8 ohms impedance

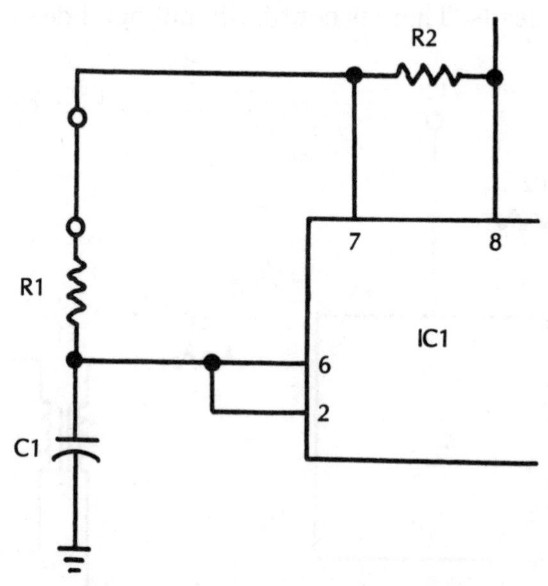

Fig. 7-2 Equivalent circuit for a dead short.

Figure 7-3 shows the effective circuit for this case. Resistor R_x is the unknown resistance being tested. This resistance is made part of the 555's timing (frequency-determining) network, along with resistors R1, R2, and capacitor C1. The higher the

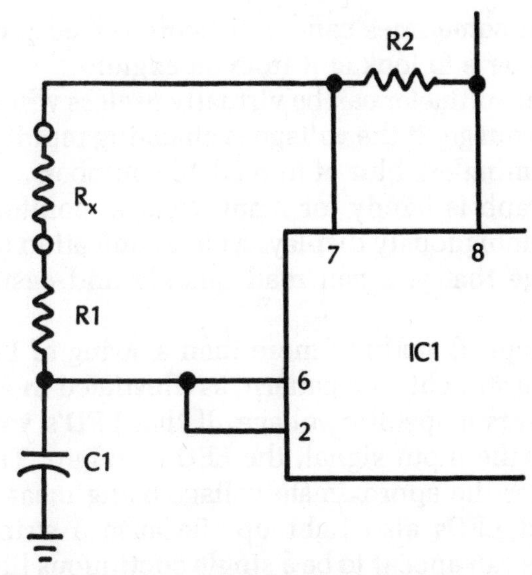

Fig. 7-3 Equivalent circuit for a finite resistance.

unknown resistance (R_x), the lower the frequency of the tone emitted by the speaker.

To change the range of the continuity tester (to change the base frequency of the oscillator), you can experiment with other values for R1, R2, and C1.

At very low frequencies, the tone from the speaker might become inaudible. You will then have to rely on the visual indication of LED D1. At extremely low frequencies (very high, measured resistances) you might be able actually to see the LED flash on and off with the low-frequency pulses generated by IC1. You can replace fixed resistor R3 with a potentiometer to provide a volume control for the speaker.

You also can use this continuity tester to check standard semiconductor junctions, such as diodes or bipolar transistors. You should hear a higher pitched tone when you connect the junction to the leads in one direction versus the other or reverse direction.

PROJECT 36: BARGRAPH VOLTMETER

There are a number of ways to obtain a voltage reading. The two most common methods are to use an analog (mechanical) meter and to use a digital readout. Both of these methods are fine for most purposes, but in some cases they are not ideal. Also, an analog meter sometimes can be difficult to read precisely, especially if you have to look at it from an angle.

A digital voltmeter can be virtually useless when monitoring a changing voltage. If the voltage is changing rapidly, the display is just a meaningless blur of unreadable numbers.

A bargraph is handy for many voltage monitoring applications. It unambiguously displays a clear indication of the approximate voltage that you can read quickly and easily, even at a distance.

A bargraph is nothing more than a string of LEDs, usually arranged in a straight-line pattern, as illustrated in Fig. 7-4. Each LED represents a specific voltage. If that LED's voltage level is exceeded by the input signal, the LED lights up. The highest lit LED indicates the approximate voltage being measured. All the lower-valued LEDs also light up. Because a string of closely spaced LEDs can appear to be a single continuous line or bar, this type of display is commonly known as a bargraph.

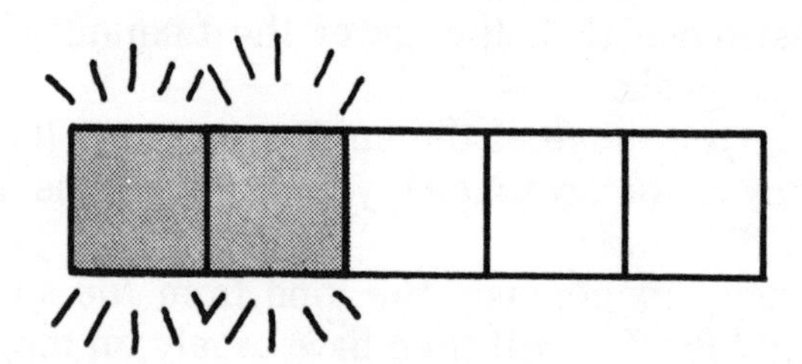

Fig. 7-4 A bargraph is made up of a string of LEDs arranged in a row pattern.

You can combine several comparator stages to build a practical bargraph-display unit. The output display (the actual bargraph) indicates the level of a voltage fed to the circuit's input. As the input voltage increases, more LEDs light up.

The schematic diagram for a simple four-stage voltage bargraph circuit is shown in Fig. 7-5. The parts list for this project appears in Table 7-2.

This circuit uses all four sections of a single LM339 quad comparator IC. If you like, you can expand the circuit easily for a wider readout range simply by adding more comparator sections. I show four stages here for no more profound reason than that there are four sections in one LM339 chip. If you use two LM339s, you can build up to an eight-stage bargraph unit. There is nothing at all essential about the number of stages.

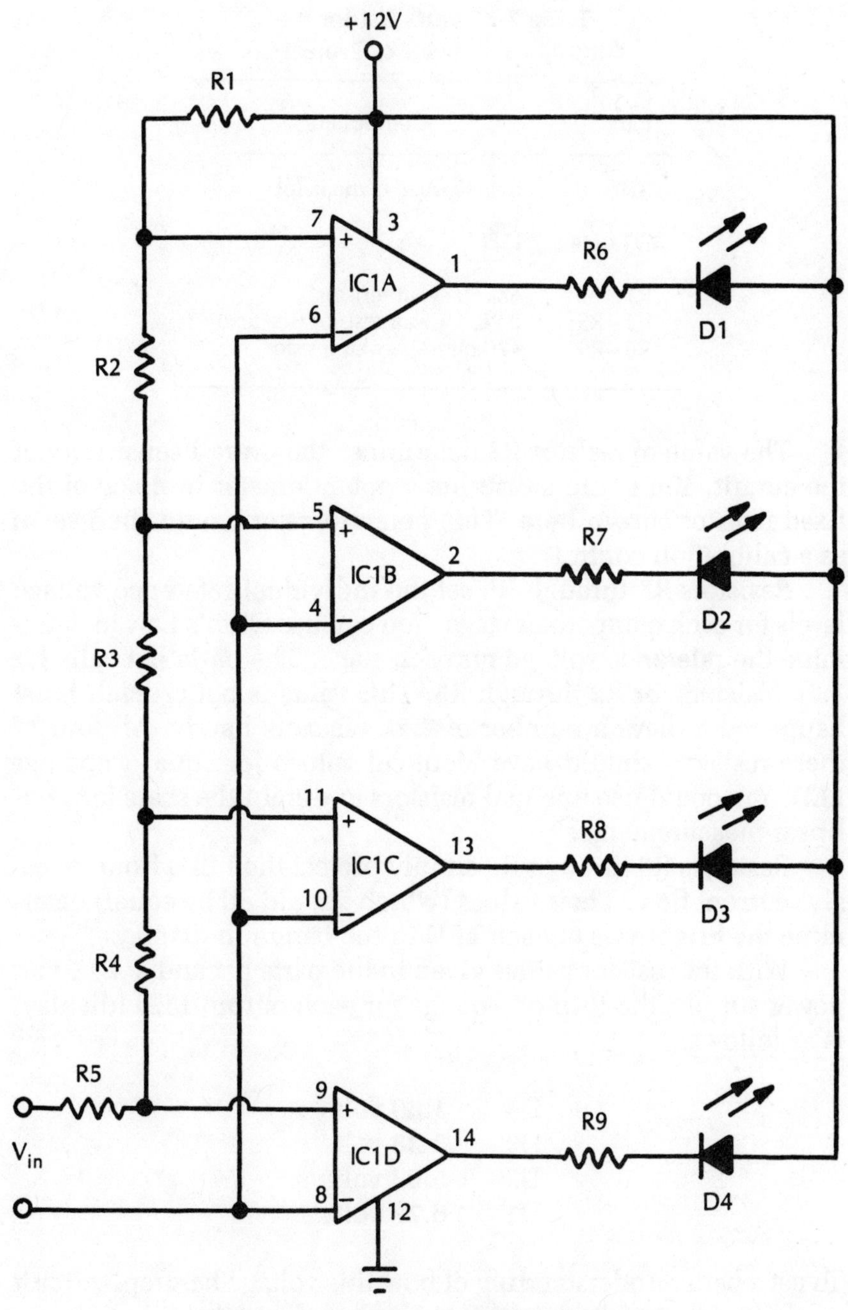

Fig. 7-5 Project 36: Bargraph Voltmeter.

Table 7-2 Parts List for the Bargraph Voltmeter of Project 36.

Part	Component
IC1	LM339 quad comparator
D1 – D4	LED
R1	68k, 1/4-watt resistor
R2 – R5	2.2k, 1/4-watt resistor (see text)
R6 – R9	470-ohm, 1/4-watt resistor

The value of resistor R1 determines the overall sensitivity of the circuit. You could substitute a potentiometer in place of the fixed resistor shown here. This potentiometer would then serve as a calibration control.

Resistors R2 through R5 set the individual reference voltage levels for each comparator stage. You can use Ohm's Law to determine the reference voltage for each stage. The parts list calls for 2.2k resistors for R2 through R5. This value is not crucial. I just happened to have a number of 2.2k resistors handy. All four of these resistors should have identical values for equal steps per LED. You could use unequal resistors to weight the scale for nonlinear measurements.

Resistors R6 through R9 simply protect the LEDs from excessive current flow. Their values (which should all be equal) determine the brightness of each LED in the bargraph display.

With the resistor values given in the parts list and a + 12-volt power supply, the turn-on voltage for each output LED (display) is as follows:

D4	1.69 volts
D3	3.38 volts
D2	5.08 volts
D1	6.77 volts

To get a better understanding of how this voltage bargraph circuit works, Table 7-3 shows a few examples of which LEDs light up for various input voltages.

Table 7-3 LEDs and Voltages for which They will Light Up.

Input Voltage	Lit LED(s)
0.00 volt	none
0.50 volt	none
1.00 volt	none
1.50 volts	D4
2.00 volts	D4
2.50 volts	D4
3.00 volts	D4
3.50 volts	D4, D3
4.00 volts	D4, D3
4.50 volts	D4, D3
5.00 volts	D4, D3
5.50 volts	D4, D3, D2
6.00 volts	D4, D3, D2
6.50 volts	D4, D3, D2
7.00 volts	D4, D3, D2, D1
7.50 volts	D4, D3, D2, D1
8.00 volts	D4, D3, D2, D1

The only other thing to bear in mind with this project is that the input signal (the voltage to be monitored) is not referenced directly to ground. You connect the input voltage across the two points so marked in the schematic diagram.

Feel free to experiment with any or all of the component values in the parts list.

PROJECT 37: SIGNAL INJECTOR

A signal injector is a handy piece of test equipment. A signal injector is used to locate the problem stage in a multistage piece of equipment. It produces a continuous signal that you can feed into the circuit under test at various points.

Starting at the input of the last stage, you inject a signal into the circuit. If the signal appears correct at the output, then that stage is all right. It is functioning correctly. The problem must be at some earlier point in the circuit.

Move the signal injector back, stage by stage, until the signal at the output disappears or becomes excessively distorted. The first stage in which this occurs has some defect. Use a multimeter or oscilloscope to locate the defective component or components.

A simple signal-injector project appears in Fig. 7-6. Table 7-4 shows the parts list.

With the component values listed in the table, the signal frequency is approximately 1 kHz (1000 Hz). You can use this circuit with the next one, the Signal-Tracer (Project 38).

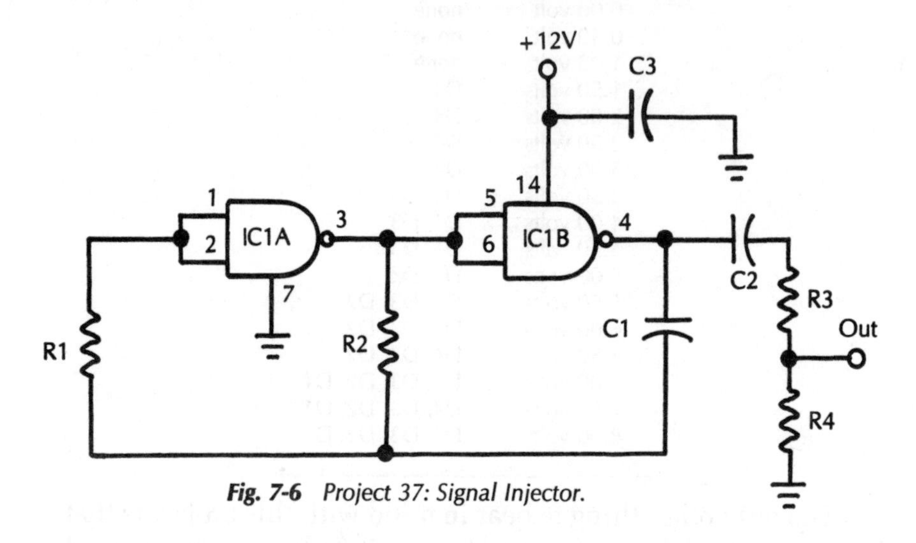

Fig. 7-6 *Project 37: Signal Injector.*

Table 7-4 Parts List for the Signal Injector of Project 37.

Part	Component
IC1	CD4011 quad NAND gate
C1	0.007-μF capacitor
C2, C3	0.1-μF capacitor
R1	220k, 1/4-watt resistor
R2	100k, 1/4-watt resistor
R3	6.8k, 1/4-watt resistor
R4	2.2k, 1/4-watt resistor

PROJECT 38: SIGNAL TRACER

A signal tracer operates the reverse of the signal injector discussed in the last section. This device detects the presence of a signal in the circuit under test. It does not produce a signal of its own.

Although you test the output stage first for a signal injector, you reverse the testing process for a signal tracer. Start at the input of the circuit under test, then move forward stage by stage until the signal disappears or becomes distorted. The problem is located in the stage immediately preceding the current test point at which the signal tracer is connected.

A signal-tracer circuit is illustrated in Fig. 7-7. A parts list for this project appears in Table 7-5.

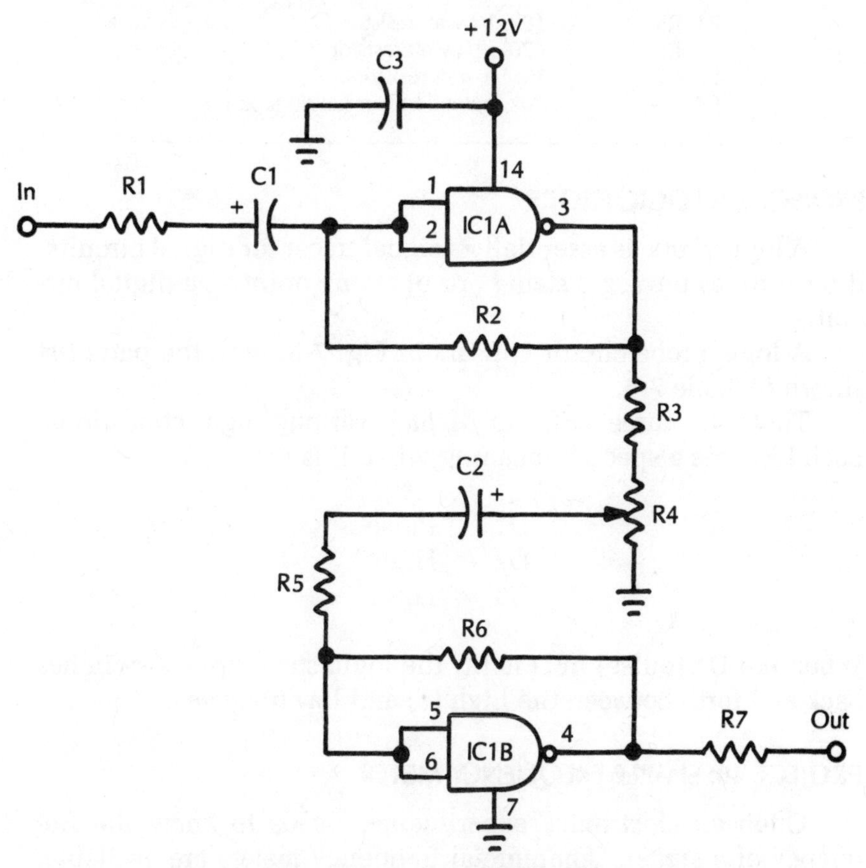

Fig. 7-7 Project 38: Signal Tracer.

Both this project and the signal-injector circuit of Project 37 utilize just one-half of a CD4011 quad NAND gate IC. You can build both into a single housing, using just a single IC chip between them. Of course, you have to change the pin numbers for one of the two circuits.

Table 7-5 Parts List for the Signal Tracer of Project 38.

Part	Component
IC1	CD4011 quad NAND gate
C1, C2	10-μF, 15-volt electrolytic capacitor
C3	0.1-μF capacitor
R1, R5	10k, 1/4-watt resistor
R2, R6, R7	470k, 1/4-watt resistor
R3	1k, 1/4-watt resistor
R4	10k, potentiometer (audio taper)

PROJECT 39: LOGIC PROBE

A logic probe is essentially a signal tracer for digital circuits. It determines the logic state (1 or 0) at any point in a digital circuit.

A logic probe circuit appears in Fig. 7-8, with the parts list shown in Table 7-6.

There are three LEDs to display various logic conditions. Each LED has a specific meaning when it is lit:

D1	Pulse
D2	High
D3	Low

When the D1 (pulse) LED is lit, the logic state rapidly switches back and forth between the high (1) and low (0) states.

PROJECT 40: SIMPLE FREQUENCY METER

Often an electronics experimenter needs to know the frequency of a signal. Commercial frequency meters are available, but they tend to be complex, expensive devices. This project is a "quick and dirty" frequency meter that will give you good results for under $10. Figure 7-9 shows the schematic diagram for this project. The complete parts list for the simple frequency-meter project appears in Table 7-7.

This circuit converts the input frequency into a proportional current that is displayed on a 1-mA current meter (M1).

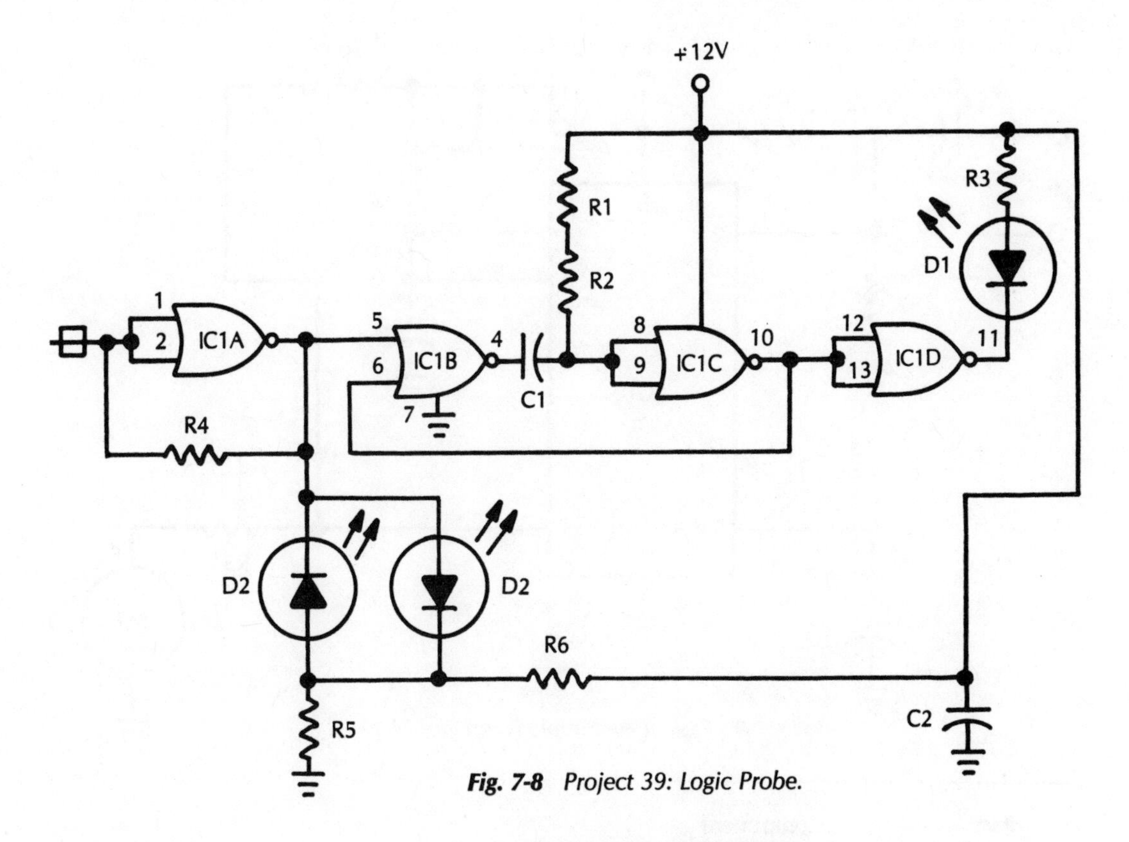

Fig. 7-8 *Project 39: Logic Probe.*

Table 7-6 Parts List for the Logic Probe of Project 39.

Part	Component
IC1	CD4001 quad NOR gate
D1 – D3	LED
C1, C2	0.1-μF capacitor
R1, R2	3.3-megohm, 1/4-watt resistor
R3, R5, R6	1k, 1/4-watt resistor
R4	2.2-megohm, 1/4-watt resistor

You use the 10k trimpot (R5) to obtain a correct zero reading on the meter. You use trimpot R4 to calibrate the meter reading for a test signal with a known frequency. These two controls interact somewhat, so take your time with the calibration procedure.

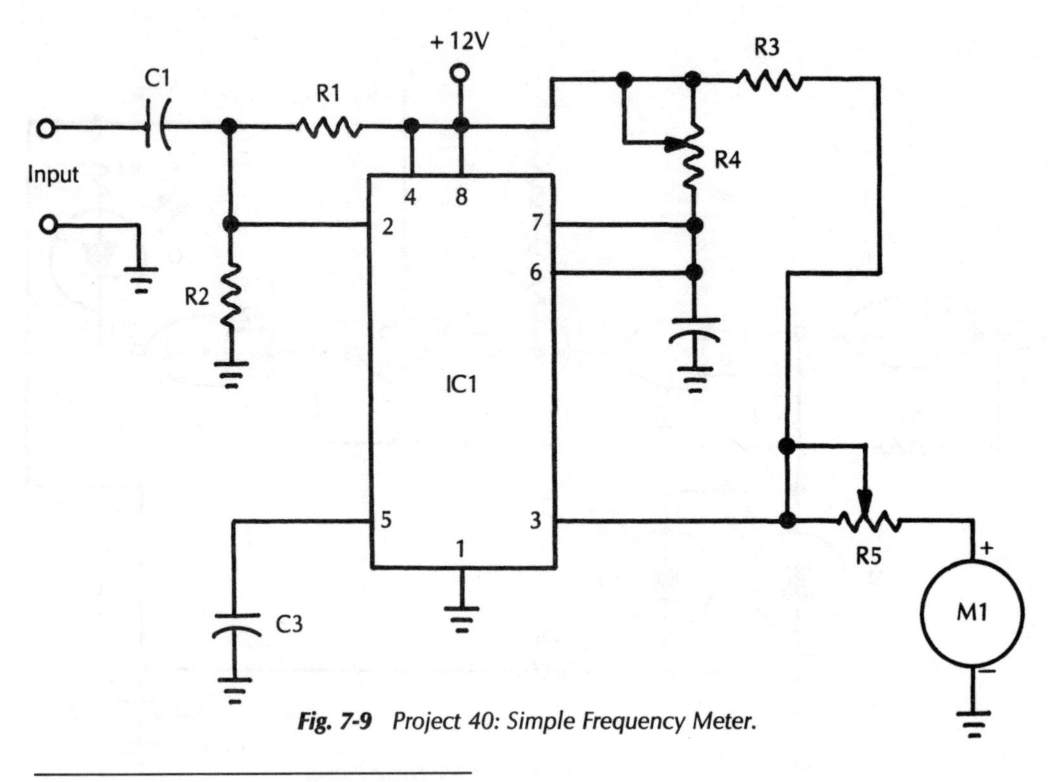

Fig. 7-9 Project 40: Simple Frequency Meter.

Table 7-7 Parts List for the Simple Frequency Meter of Project 40.

Part	Component
IC1	7555 (or 555) timer
C1, C3	0.01-μF capacitor
C2	0.22-μF capacitor
R1, R2, R3	4.7k, 1/4-watt resistor
R4	250k trimpot
R5	10k trimpot
M1	1-mA meter

The primary restriction for this frequency meter is that the input signal must be a good approximation to a square wave for reliable operation. Other input waveforms might not give accurate results. To make a more versatile piece of test equipment, you could add a Schmitt trigger stage to the input of the circuit shown here. This will convert almost any waveform into a quasi-square wave that should be readable by the circuit.

PROJECT 41: AUDIO-FREQUENCY METER

In electronics work, you occasionally need to determine the frequency of an unknown ac signal. There are several ways to do this. You can use a digital-frequency meter, but commercial units are fairly expensive, and as a project, a digital-frequency meter is fairly complex.

Generally the simplest and least expensive method of frequency measurement is to use a simple analog-frequency meter. I described one such circuit in Project 40. Figure 7-10 shows an alternate and slightly more accurate frequency-meter circuit designed specifically for use with audio frequencies. A parts list for this project appears in Table 7-8.

Basically, this circuit is a frequency-to-current converter. The output current is displayed on a small panel meter (M1). This current is directly proportional to the input frequency.

A circuit like this operates best if the input is a simple (low-harmonic content) waveform, preferably a sine wave. It might be confused by strong harmonics in certain complex signals.

This project is specifically designed to function with frequencies in the audio range. As a rule of thumb, the audio band is considered to extend from approximately 20 Hz up to about 20 kHz (20,000 Hz).

Fig. 7-10 *Project 41: Audio-Frequency Meter.*

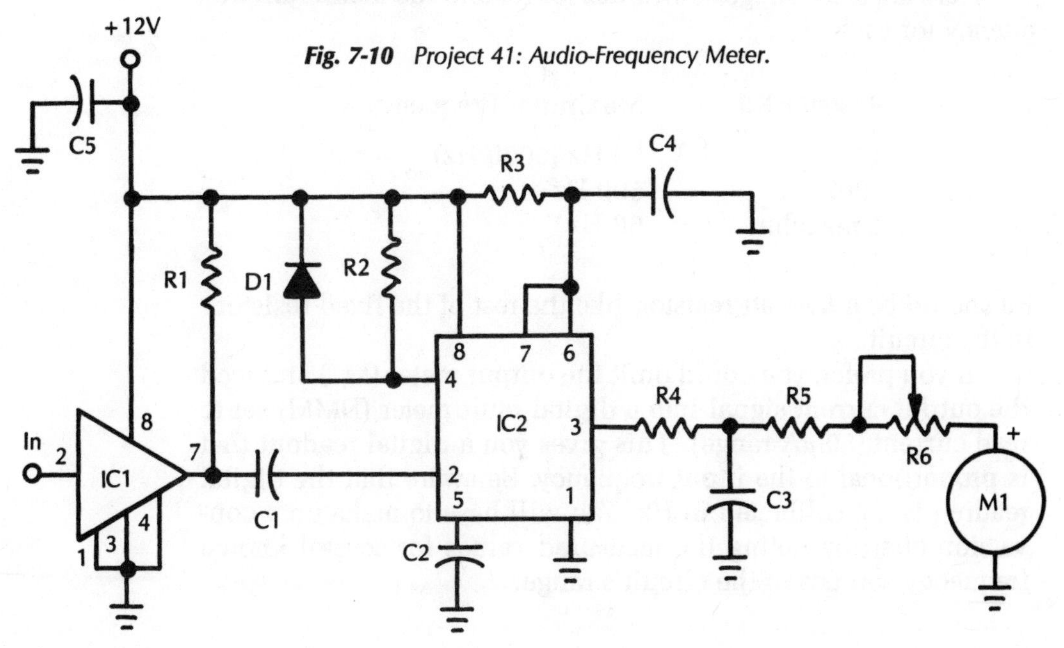

Table 7-8 Parts List for the Audio-Frequency Meter of Project 41.

Part	Component
IC1	311 amplifier
IC2	555 timer
D1	Diode (1N914, or similar)
C1	500-pF capacitor
C2	0.01-μF capacitor
C3	8.2-pF capacitor
C4	0.01-μF, silver mica capacitor
R1,R2,R5	1k, 1/4-watt resistor
R3	See text
R4	56k, 1/4-watt resistor
R6	250k potentiometer
M1	50-μA meter

Resistor R3 determines the exact range of this circuit. The higher the resistance of this component, the lower the input frequency for a full-scale reading on the meter.

Here are three suggested values for R3 and the maximum frequency for each:

Resistor R3	Maximum Frequency
10k	5 kHz (5000 Hz)
100k	500 Hz
1 megohm	50 Hz

R3 should be a 1/2-watt resistor, like the rest of the fixed resistors in the circuit.

If you prefer, you could omit the output meter (M1) and feed the output current signal into a digital multimeter (DMM) set to read current (50-μA range). This gives you a digital readout that is proportional to the input frequency. Be aware that the digital reading is not calibrated in Hz. You will have to make up a conversion chart by noting the measured values for several known frequency sources in the circuit's range.

PROJECT 42: DIGITAL-FREQUENCY METER

Projects 40 and 41 are analog frequency meters. The circuits are somewhat crude, and it is difficult to get a precise reading. They are fine for some applications (in which only approximate readings are required). For more critical applications, a digital-frequency meter is highly desirable. This project is just such a device. Before describing the project itself, let's take a moment to consider just how digital-frequency meters work in principle.

There are several possible approaches to digital-frequency measurement. The most common system today is the *window-counting method.* A sample of the input signal (the frequency to be measured) is allowed to pass through a gate. This sample lasts for a specific and fixed period of time. By counting the number of pulses that get through the gate during this sample period, you can calculate the input frequency with reasonably high accuracy.

Figure 7-11 shows a block diagram for a typical window-type digital-frequency meter. The input signal is fed first through an amplifier that functions as a buffer and (in some designs) boosts the input signal up to a usable level. This preamplification stage is not always included in all digital-frequency-meter circuits. It is included in most better devices of this type. The preamplifier stage improves the sensitivity of the instrument, permitting accurate measurement of relatively low-level signals.

The next stage in the block diagram is a Schmitt trigger that converts (almost) any input waveshape into a good enough rectangle wave to be recognized reliably by the digital circuitry. If you are only measuring square-, rectangle-, or pulse-wave signals with the frequency meter, you can omit Schmitt trigger from the circuit.

Next, the processed input signal from the Schmitt trigger is fed to one input of an AND gate. The other input to the AND gate is a regular stream of pulses from an internal reference oscillator. This signal generally is called a *time base* for this type of application. The time base (oscillator) stage puts out three synchronized signals (or a single signal tapped off with delay circuits, as shown in the diagram). The timing (phase) relationships of the three time-base signals (illustrated in Fig. 7-12) are crucial.

The first of these time-base signals (labelled "Gate" in Fig. 7-12) is fed to the second (control) input of the AND gate. This

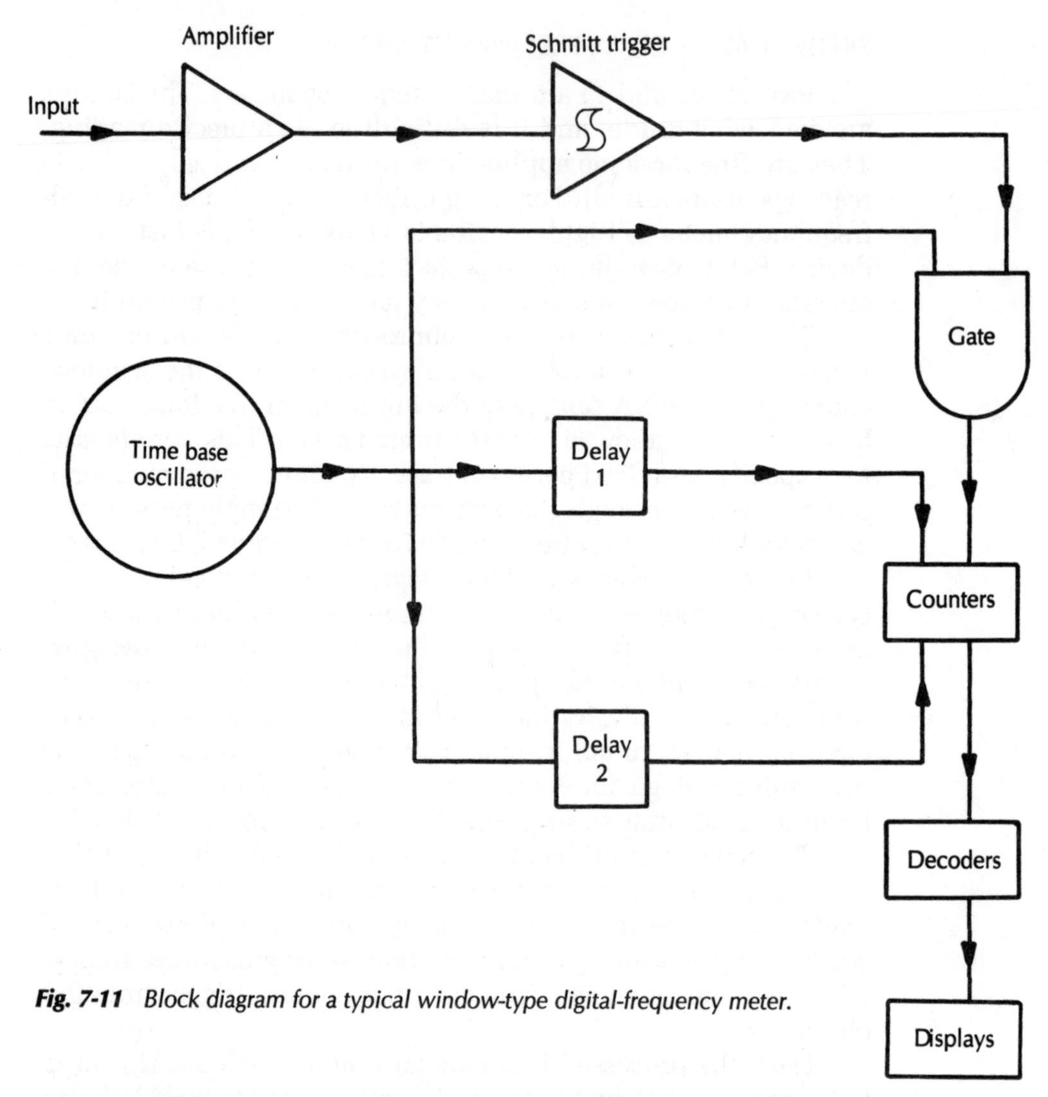

Fig. 7-11 *Block diagram for a typical window-type digital-frequency meter.*

signal effectively "opens the window" when it is at logic 1 (high) and "closes the window" when it is at logic 0 (low).

The second time-base signal ("Latch") is delayed until after the first has dropped back to its normal low state. This signal latches the output of the counters so that they can hold their final value long enough to produce a readable display.

The third and last of the time-base signals (Reset) goes high after the Latch signal has returned to low. This signal is used to reset the counters back to zero in preparation for the next measurement cycle (triggered when Gate goes high again).

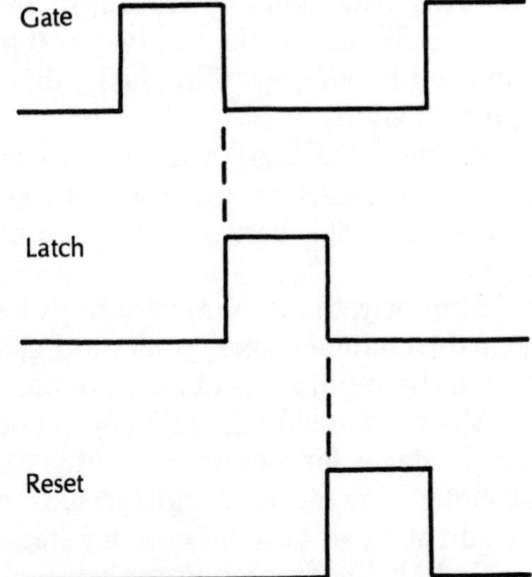

Fig. 7-12 *The phase relationships for the three time-base signals are crucial.*

The output from the AND gate is counted by the counter stage under control of the Latch and Reset signals, as discussed above.

The accuracy of a digital-frequency meter of this type is generally given as x percent ± one digit. *X* is the overall accuracy of the measurement circuit. The other half of the accuracy rating (± one digit) indicates that the least significant digit in the display might bob up and down between adjacent values on successive measurements. This occurs whenever a partial input pulse happens to get through the window. For example, if 573.5 input pulses are passed during one Gate cycle, on some cycles the output count will be 573 Hz, while other measurement cycles, the count will reach 574 Hz.

For a functional digital-frequency meter, the time-base oscillator must be extremely precise and stable in its output frequency. There must be an absolute minimum of frequency drift in the oscillator circuit. Any error here changes the length of the window period. If the window is too long, too many input pulses pass through to the counter. If the window period is too short, too few input pulses reach the counter stages. Any error in the time-base frequency inevitably throws off the output readings (display), often by a significant amount. In many frequency-counter circuits, a crystal oscillator provides the time-base signals.

The input frequency must be higher than the reference

(time-base) frequency. If the input frequency is lower than the reference frequency, then only 1 or 0 pulses can ever get through each sample window. Obviously, this does not provide any useful information.

If you must use your digital-frequency meter to measure lower frequencies, you can add an additional frequency-multiplier input stage between the Schmitt trigger and the gating stage.

Similarly, to measure very high frequencies that might overrange the counter stages, you could use a frequency-divider stage to drop the input signal to a more convenient lower frequency.

Most commercial frequency counters have three to six counter stages for maximum counts of 999 to 999,999. The project shown here has four digits (maximum count of 9999), but you can add some additional counter stages if you prefer.

Switchable frequency multipliers and/or dividers often are included also in commercial units to allow for manually selectable ranges. Movable decimal points might or might not be included in the display readout. There is no reason why you can't add these modifications easily to this project. I present the digital-frequency meter project in its basic form here. Extra features are optional. I leave the design of the additional circuitry for the suggested modifications to you.

Figure 7-13 shows the schematic for a practical digital-frequency-meter circuit. The parts list for this project appears in Table 7-9.

Each digit of the output display is driven by its own individual CD4026 decade counter (IC2 through IC5). Four digits are shown here. To add more digits to the readout, simply connect pin 5 from the last stage to pin 1 of the next stage. You connect the other pins for each additional counter IC in exactly the same way as shown for IC2 through IC5. You can expand the display readout to include as many digits as you like, although the law of diminishing returns starts to set in if you use more than about six digits.

Transistor Q1 and IC1 precondition the input signal as described in the general circuit analysis above. This stage of the circuit ensures that the input signal has an acceptable level and waveshape to be counted reliably by the digital circuits in later stages.

IC6 serves as the reference oscillator, or time base, for the project. You can fine-tune the time-base frequency with R33, a 1-megohm trimpot. You can calibrate this circuit by simply monitoring a known frequency source (a calibrated signal generator is best) while adjusting R33 to obtain the correct value on the circuits. After calibrating the digital-frequency meter, do not change the setting of trimpot R33. You might want to place a drop of paint or glue on the trimpot to keep it from mechanically drifting out of position.

PROJECT 43: DIGITAL-CAPACITANCE METER

It is quite easy to measure resistance. An ohmmeter is a fairly simple device. This is fortunate because the most common type of electronic component is the resistor.

Unfortunately, the second most common type of electronic component is the capacitor. It is not so easy to measure capacitance, because it is an ac rather than a dc phenomenon.

Before the advent of digital ICs, a number of analog-capacitance-meter circuits were designed, but they tended to be complex, expensive, and not particularly reliable or accurate, especially for low values of capacitance.

Today digital circuitry places efficient, convenient, and reasonably accurate capacitance measurements within reach of the average electronics hobbyist, without an exorbitant price tag. Capacitance values can be measured easily over a wide range with the use of digital ICs. The basic circuitry is not very different from that found in digital voltmeters. Figure 7-14 shows a block diagram of a basic digital-capacitance meter.

In both analog and digital-capacitance meters, the basic idea is to measure the time it takes the unknown capacitance to charge (to $^2/_3$ of its full capacity) through a known resistance. This is far easier to accomplish with digital circuitry than with analog circuitry.

Take a look at the block diagram in Fig. 7-14. The first stage is a simple *monostable multivibrator*, or *one-shot*. A monostable multivibrator has one stable output state. Its output holds that stable state indefinitely until the input receives another trigger pulse. The output then jumps to the opposite (nonstable) state for a fixed period of time. This time period is determined by component values in the multivibrator circuit. Specifically, the resistance and capacitance control the time period. Altering either

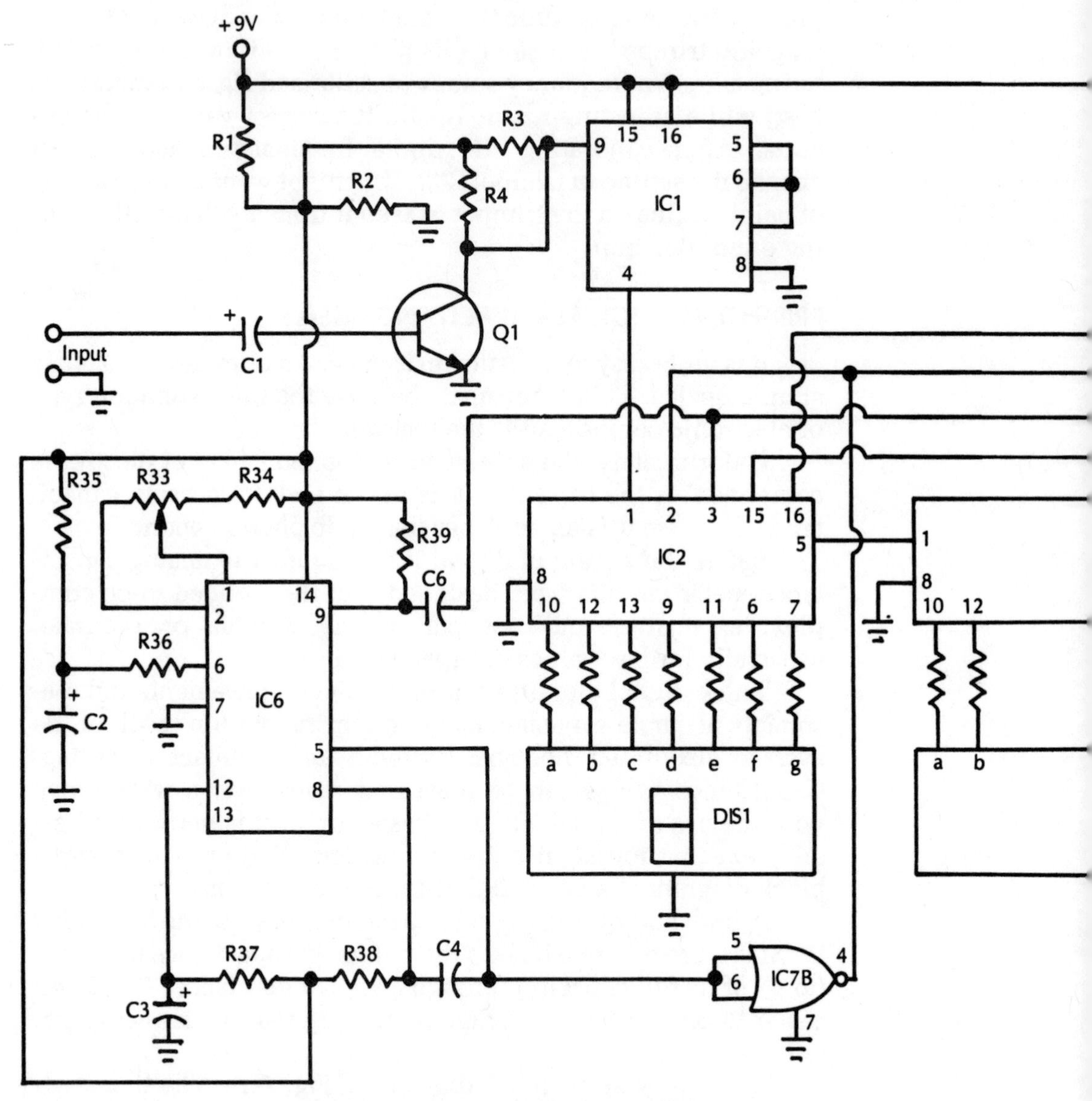

Fig. 7-13 Project 42: Digital-Frequency Meter.

value changes the time period of the multivibrator. When the multivibrator's time period runs out, the output returns to the original, stable state until another trigger pulse is received.

A monostable multivibrator can be stable either high or low. For convenience in this discussion, I will assume that the stable state is a logic 0 (low).

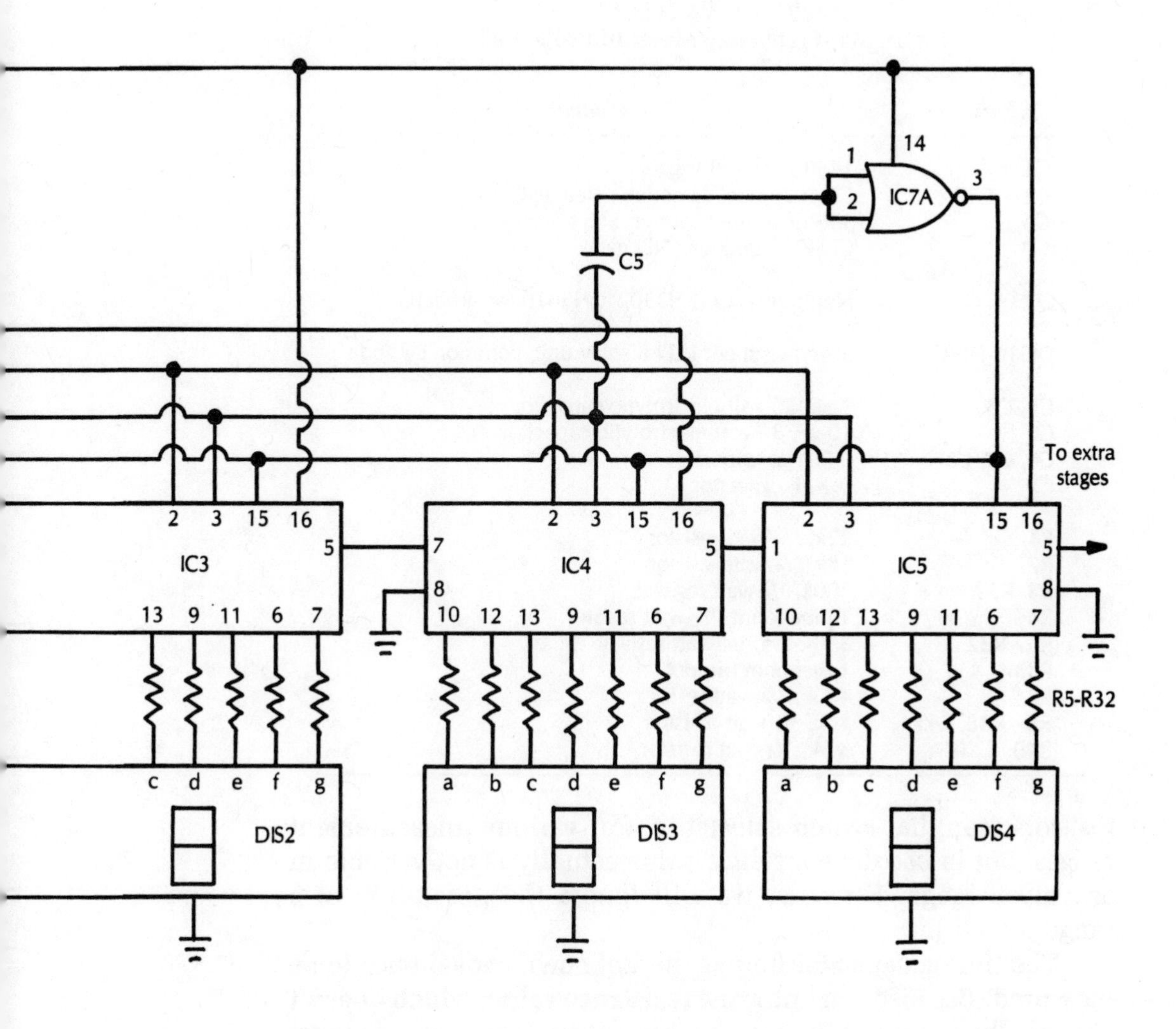

The output of the monostable multivibrator stage remains at logic 0 until the circuit is triggered. At that instant, the output of this stage switches to logic 1 for a specific period of time that is determined by the values of the resistor/capacitor combination. In this application, the timing resistor has a fixed (known) value. (In some practical digital-capacitance meters, different timing

Table 7-9 Parts List for the Digital-Frequency Meter of Project 42.

Part	Component
IC1	14583 Schmitt trigger
IC2 – IC5	CD4026 decade counter (see text)
IC6	556 dual timer (or two 555 timers)
IC7	CD4011 quad NAND gate
Q1	Npn transistor (2N3302, 2N5826, or similar)
DIS1 – DIS4	Seven-segment LED display unit, common cathode
C1, C3	1-μF, 30-volt electrolytic capacitor
C2	10-μF, 30-volt electrolytic capacitor
C4, C5, C6	0.001-μF capacitor
C7	0.1-μF capacitor
R1	22k, 1/4-watt resistor
R2	18k, 1/4-watt resistor
R3, R39	100k, 1/4-watt resistor
R4	10-megohm, 1/4-watt resistor
R5 – R32	330-ohm, 1/4-watt resistor
R33	1-megohm trimpot
R34	470k, 1/4-watt resistor
R5 – R38, R41	10k, 1/4-watt resistor
R40	3.3k, 1/4-watt resistor

resistors can be switch-selectable for various measurement ranges, but in use the resistance value actually is not variable in any given range. For now, we will ignore the possibility of a range switch.).

The timing capacitor here is the unknown capacitance to be measured. Because you know the resistance value, which doesn't change, the output of the monostable multivibrator goes to logic 1 when triggered for a period of time that is directly proportional to the value of the unknown input capacitance.

In contrast to a monostable multivibrator, an *oscillator*, or *clock stage*, generates pulses at a known (high-frequency) rate.

The output of the monostable multivibrator controls an AND gate. When the multivibrator is putting out a logic 1 (triggered, nonstable state), the oscillator pulses can get through the gate. When the multivibrator's output is logic 0 (stable state), the clock pulses are blocked.

The pulses that are permitted to pass through the gate are

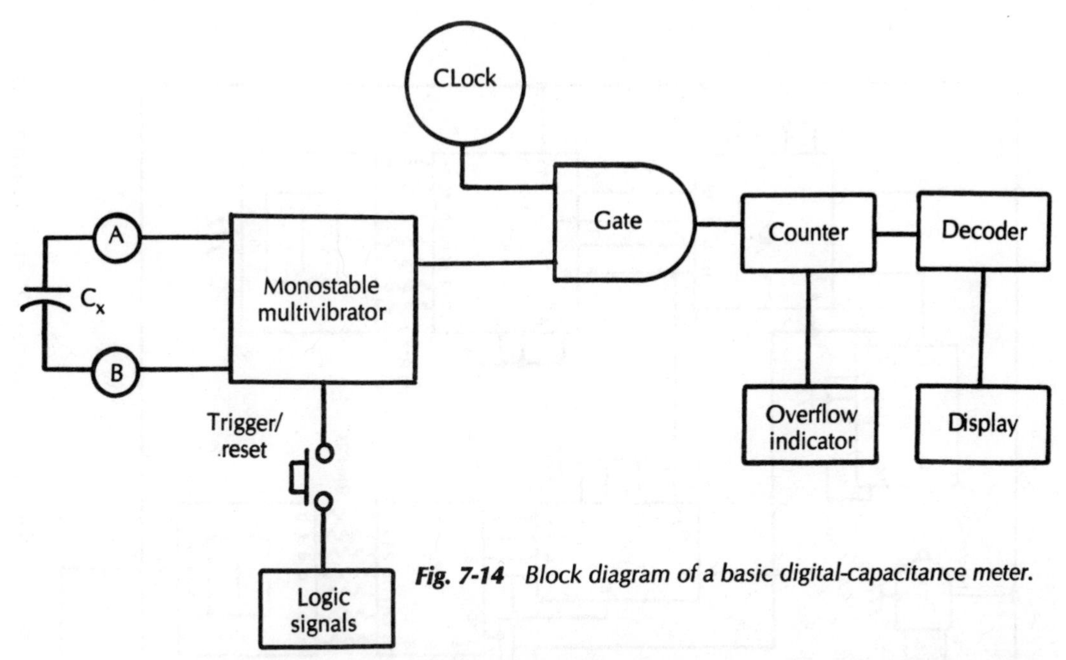

Fig. 7-14 Block diagram of a basic digital-capacitance meter.

counted, and the total number (which will be directly proportional to the unknown capacitance) is displayed on the readout.

You use a pushbutton switch to reset the counter stages and to trigger the monostable multivibrator manually when you want to take a new measurement. You should mount this switch on the front panel of the instrument for easy access.

Figure 7-15 shows a practical digital-capacitance-meter circuit. The parts list appears in Table 7-10.

This project is capable of measuring capacitances ranging from less than 100pF to well over 1000μF. The vast majority of capacitors you are likely to work with as an electronics hobbyist will probably fall within this range.

If you are going to test electrolytic capacitors or other polarized components with this circuit, be sure to hook up the meter with the correct polarity. You should attach point A to the capacitor's positive lead. Point B is the negative connection point or ground.

Resistor R1 sets the full-scale reading for the meter. This component should be a trimpot (preferably a ten-turn type) that is set carefully with a high-grade (precision) capacitor with a known value hooked up to the test leads. Once you have set the desired resistance, you should leave the trimpot strictly alone. It

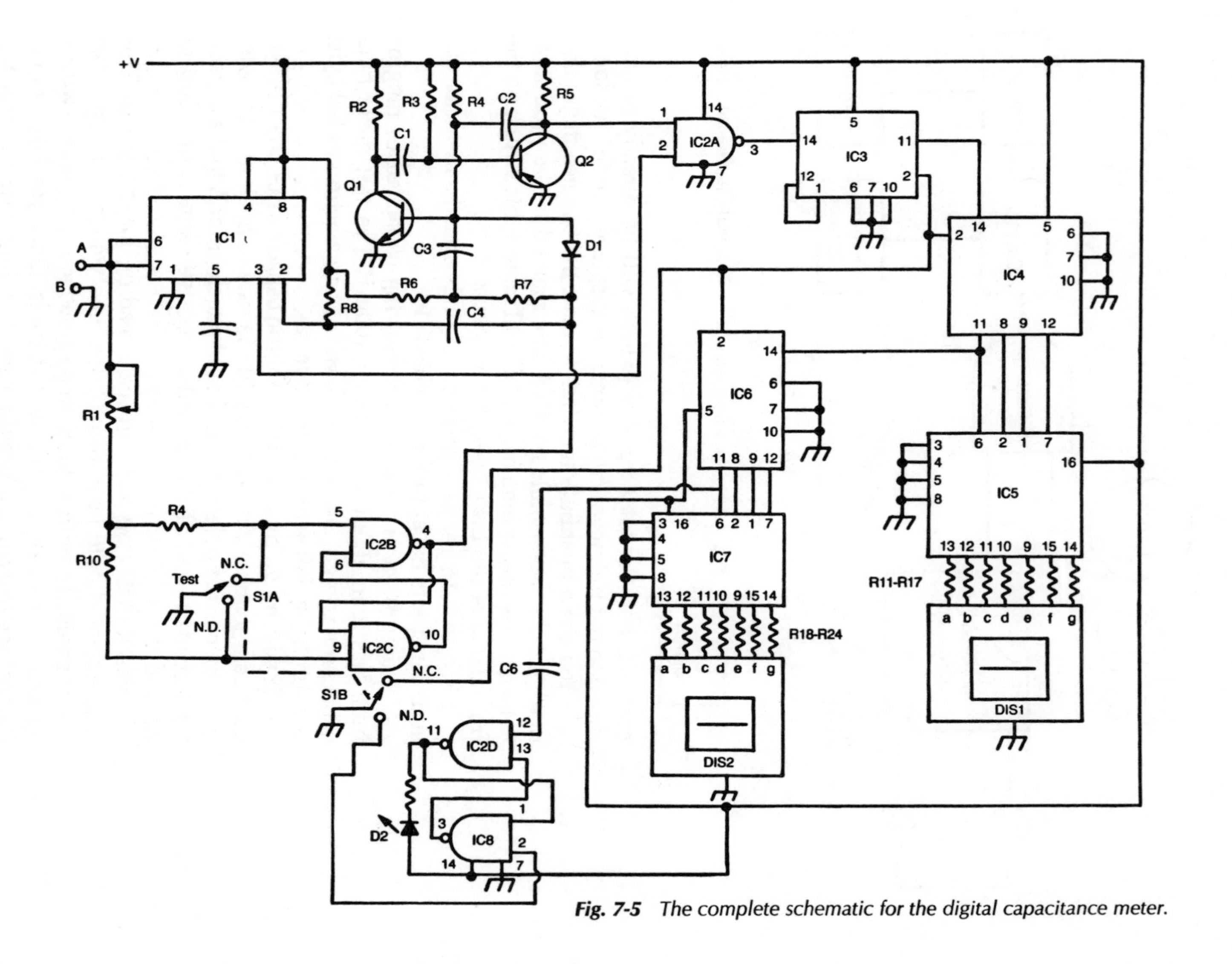

Fig. 7-5 *The complete schematic for the digital capacitance meter.*

Table 7-10 Parts List for the Digital-Capacitance Meter of Project 43.

Part	Component
IC1	7555 timer (or 555)
IC2, IC8	CD4011 quad NAND gate
IC3, IC4, IC6	74C90 decade counter
IC5, IC7	CD4511 BCD-to-7-segment decoder
Q1, Q2	Npn transistor (2N3907, or similar)
D1	Diode (1N4734, or similar)
D2	Red LED (overflow indicator)
DIS1, DIS2	Seven-segment LED display unit, common cathode
C1, C2	0.047-μF capacitor
C3, C7	0.1-μF capacitor
C4, C5	0.01-μF capacitor
C6	0.0022-μF capacitor
R1	Calibration trimpot/resistor (see text)
R2, R5	2.7k, 1/4-watt resistor
R3, R4, R7	15k, 1/4-watt resistor
R6, R8, R26	10k, 1/4-watt resistor
R9, R10	1.8k, 1/4-watt resistor
R11 – R24	330-ohm, 1/4-watt resistor
R25	3.3k, 1/4-watt resistor
S1	DPDT pushbutton switch, normally open (push to clear and test—see text)

should be mounted in a relatively inaccessible position. You might want to apply a dab of paint to prevent the trimpot's slider from changing position. Even better, after you have calibrated the unit, carefully remove the trimpot from the circuit without letting its setting change at all. Measure the resistance and replace the trimpot with an appropriately valued resistor (1 percent tolerance, or better).

The value of R1 also determines the measurement range. You should use smaller resistance values to measure larger capacitances. A 100k trimpot is a good starting point for a ×1 scale, while a 10k scale is better served by a 5k trimpot. This is why no specific value for R1 is given in the parts list.

For maximum versatility, duplicate R1 to create overlapping ranges, and use a rotary switch to select the appropriate resistor for each measurement.

8 ❖ Miscellaneous Circuits

IN WRITING A BOOK OF PROJECTS LIKE THIS, ORGANIZATION CAN sometimes be a problem. I have tried to group the projects into logical chapter categories. In some cases, there is an overlap between categories. For example, the Emergency-Flasher Project, Project 1, appears in chapter 2 (Automotive Projects), but it also could fit into chapter 5 (Lamps and Displays). Project 14 is a car clock. It appears in chapter 3 (Timing Circuits), but it also could fit logically into chapter 2 (Automotive Projects).

Other projects don't quite fit into any of the main chapter categories. Rather than include several chapters of just one or two projects each, I am grouping these "stray" projects together as miscellaneous.

PROJECT 44: LIQUID DETECTOR

Have you ever had a problem with flooding in your basement? Is there an area you must keep dry? Do you need to know when a liquid level reaches a certain point?

If you answer "yes" to any of these questions, you will appreciate this project. This circuit can alert you when it detects liquid at a specific level.

The schematic diagram for this simple, but useful, project appears in Fig. 8-1. Table 8-1 is the parts list for this project.

You need to place two probes at the location to be monitored. If there is no liquid at this point, there will be an open circuit between the two probes. When the water (or other liquid) level rises enough to come into contact with both probes, however, it creates a short circuit between them. The relay then activates.

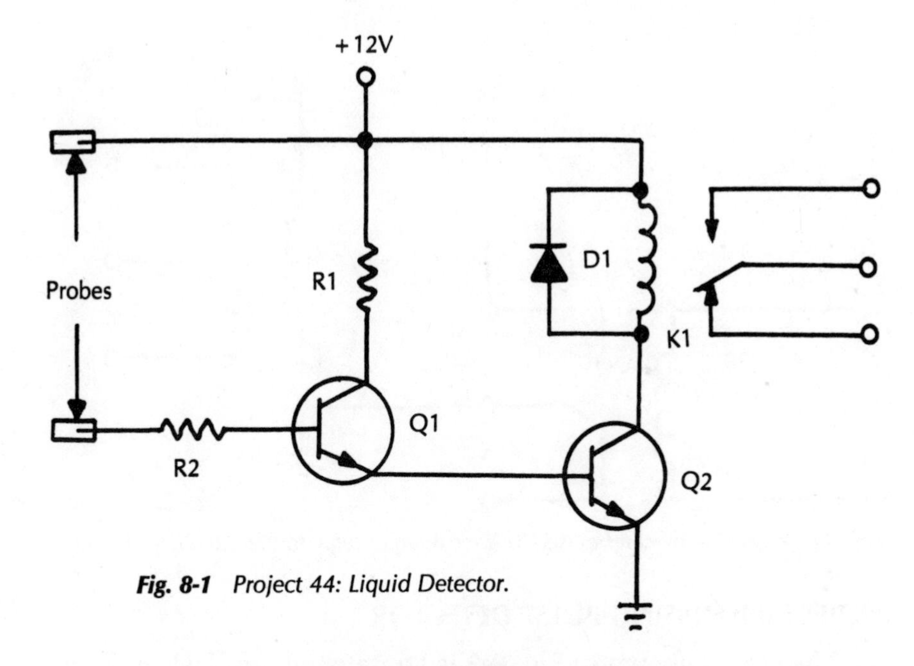

Fig. 8-1 *Project 44: Liquid Detector.*

Table 8-1 Parts List for the Liquid Detector of Project 44.

Part	Component
Q1, Q2	Npn transistor (2N3904, or similar)
D1	Diode (1N914, or similar)
R1, R2	3.3k, 1/4-watt resistor
K1	12-volt relay with contacts to suit desired load

The relay can control an alarm device or perhaps a pump. If your application involves filling a vessel up to a specific level (monitored by the probes), the relay can control the valve, permitting the liquid to flow into the vessel. In this case, you would use the relay's normally closed contacts, so the valve turns off when the liquid shorts out the probes.

Of course, you should select the relay so that its contacts can safely handle the current drawn by the desired load device. If you are driving a very large load, you might need to use the primary relay to drive a second, larger relay, as illustrated in Fig. 8-2.

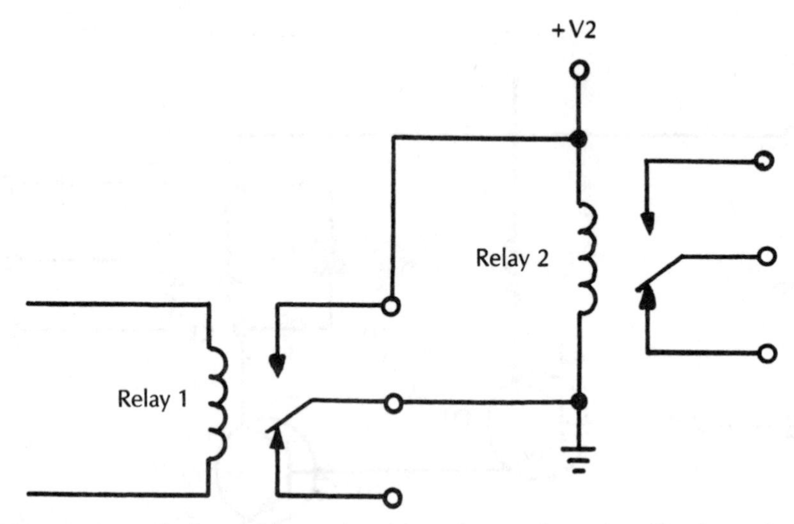

Fig. 8-2 *A small relay can be used to drive a larger relay to handle a heavy load.*

PROJECT 45: MISSING-PULSE DETECTOR

The circuit shown in Fig. 8-3 is an unusual application for a 7555 (or 555) timer IC. Table 8-2 shows the parts list. This is a missing-pulse detector. The circuit produces a negative-going pulse when it detects a gap (or missing pulse) in a stream of pulses fed to the input. Missing-pulse detectors like this typically are used in continuity testers and security-alarm systems, among other applications.

Basically, this circuit is a monostable multivibrator. Normally, if there is no gap in the pulse stream, the multivibrator is retriggered continually by new pulses at its input before it has a chance to time out. Transistor Q1 adds this retriggering characteristic to the basic timer.

Ordinarily then, a new pulse is received at the input before the timer has finished cycling through its time period. Each new input pulse initiates a new timing cycle. This can continue indefinitely as long as there is a steady, uninterrupted stream of input pulses. The timer never gets a chance to time out. The output remains in a constant high state.

Now suppose one (or more) of the incoming pulses is missing from the stream for some reason. The timer now has a chance to time out because it is not retriggered. The output goes low until the pulse stream at the input starts up again.

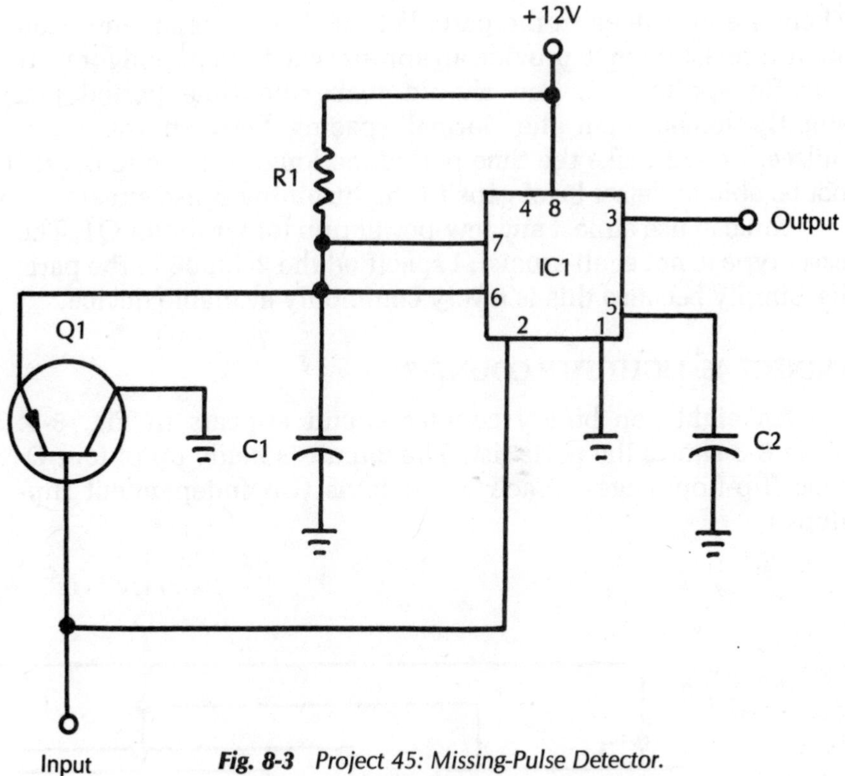

Fig. 8-3 Project 45: Missing-Pulse Detector.

Table 8-2 Parts List for the Missing-Pulse Detector of Project 45.

Part	Component
IC1	7555 timer (or 555)
Q1	Pnp transistor (2N3906, or similar)
C1	Select for desired frequency—see text
C2	0.01-μF capacitor
R1	Select for desired frequency—see text

External circuitry can easily detect this change from high to low. You can trigger almost anything by this event.

Resistor R1 and capacitor C1 set up the timing period for the timer. The formula is the usual one for 555 circuits:

$$t = 1.1\,R1C1$$

There are no values in the parts list for these components. You should select them to provide an appropriate time period for your specific application. You should make the time period just slightly longer than the normal spacing between the input pulses. Do not make the time period too long, or the circuit will not be able to detect brief gaps in the incoming pulse stream.

You can use almost any low-power pnp transistor for Q1. The exact type is not at all crucial. I specified the 2N3906 in the parts list simply because this is a very commonly available device.

PROJECT 46: EIGHT-STEP COUNTER

An eight-step binary counter circuit appears in Fig. 8-4. Table 8-3 shows the parts list. The circuit is made up of four D-type flip-flop stages. (Each IC contains two independent flip-flops.)

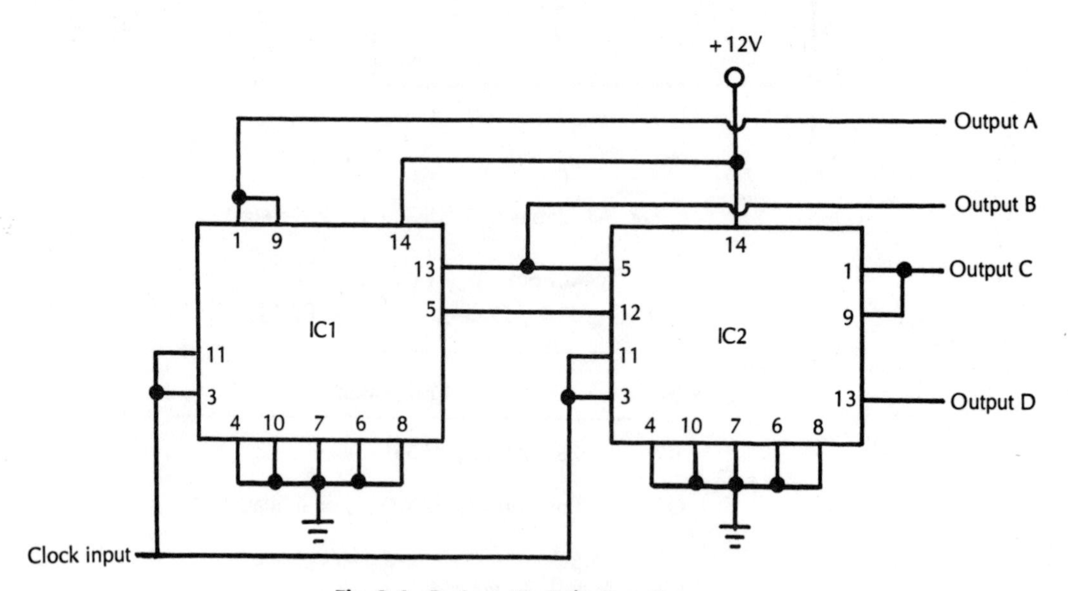

Fig. 8-4 Project 46: Eight-Step Counter.

Table 8-3 Parts List for the Eight-Step Counter of Project 46.

Part	Component
IC1, IC2	CD4013 dual D-type flip-flop

This is certainly the shortest parts list in this book. Nothing is require other than the two ICs.

The counter circuit uses the binary numbering system—that is, there are just two possible digits—0 and 1. There are four output lines, each representing a single binary digit. Thus, the total output is a four-bit binary number, or *nibble*.

The count begins at zero (0000). It advances one count each time an external trigger or clock pulse is received at the input. The counter counts up to eight (1000), then automatically resets back to zero (0000) and starts the counting cycle over.

Output A is the least significant digit (LSD), and output D is the most significant digit (MSD). Table 8-4 summarizes the counting cycle. This pattern continues to cycle through over and over, as long as trigger or clock pulses are applied to the input.

You can use the outputs to drive LEDs or almost anything else that can be controlled by a digital signal. If your intended load draws a significant amount of current, you will need to add amplifier stages or relays to the outputs.

Table 8-4 Counting Cycle for Eight-Step Counter.

Binary Count D C B A	Decimal Count
0 0 0 0	0
0 0 0 1	1
0 0 1 0	2
0 0 1 1	3
0 1 0 0	4
0 1 0 1	5
0 1 1 0	6
0 1 1 1	7
1 0 0 0	8
0 0 0 0	0
	(Note that the counter resets itself here.)
0 0 0 1	1
0 0 1 0	2
0 0 1 1	3
0 1 0 0	4
0 1 0 1	5

PROJECT 47: MAJORITY-LOGIC CIRCUIT

In most digital-gating applications, the output state is determined by very specific patterns of inputs. In some cases, however, the exact patterns of individual inputs being at specific states is not as important as broader, more generalized patterns.

For example, consider a voting-machine network. There are multiple, individually controllable inputs that are all given equal weighting, that is, any one input has the same significance as any other one input. The actual state of input C, for example, isn't important. You are interested only in which input state is dominant over the entire set of inputs. Are there more 0s or more 1s? In other words, rather than direct, fixed gating logic, you need a system of "majority rule." Not surprisingly, digital circuits that work along these lines are known as *majority-logic circuits*.

Dedicated majority-logic chips are available, but they are still pretty difficult to locate, and rather expensive. Also, it is more educational to use standard digital-gate devices, so the hobbyist obtains a clearer idea of what is happening within the circuit. The majority-logic project uses this approach.

Although this project is just a demonstration device, you can put this majority-logic circuit to work in many practical applications.

A majority-logic-gating circuit allows the inputs to "vote" on the desired output condition. It is a form of "digital democracy."

In this discussion, I assume that the majority-logic circuit is a noninverting type. Inverting majority-logic circuits are also possible. They work the same way, but the output is inverted. For example, in an inverted majority-logic circuit, a 0 input is a "vote" for a 1 output.

In a noninverted majority-logic circuit, the output state is the same as the majority of the inputs. That is, if the majority of inputs are high, the output will be high. If the majority of inputs are low, then the output will be low. To avoid problematic tie votes, there should be an odd number of inputs. That way, one input state or the other is always dominant. The smallest practical number of inputs to a majority-logic-gating circuit is three, for reasons that should be perfectly clear.

In a three-input majority-logic circuit, for the output to be high, at least two of the inputs must be high. The same is true for low signals.

Note that for a three-input gate, there are eight possible input combinations:

0 0 0
0 0 1
0 1 0
0 1 1
1 0 0
1 0 1
1 1 0
1 1 1

In majority-logic, some of these combinations are equivalent, effectively reducing the total to just four possibilities:

1. All three inputs low: 000
2. Any two inputs low: 001, 010, or 100
3. Any two inputs high: 011, 101, or 110
4. All three inputs high: 111

Note that all eight possible bit combinations are accounted for here.

Actually, you can reduce the possible significant combinations to just two. If all three inputs are the same, then at least two must be identical. In other words, either two or more inputs are low: 000, 001, 010, or 100, or two or more inputs are high: 011, 101, 110, or 111. One (and only one) of these two statements can be true for any possible input bit combination.

In truth table form, a three-input majority-logic gate performs like this:

Inputs			Output
A	B	C	
0	0	0	0
0	0	1	0
0	1	0	0

0	1	1	1
1	0	0	0
1	0	1	1
1	1	0	1
1	1	1	1

You can generate any digital input/output pattern by combining standard logic gates, so there is no reason at all why you can't create a majority-logic circuit from scratch. As is always the case with digital-gating networks, there are many possible ways to generate the desired input/output pattern. The gating circuit shown in Fig. 8-5 is probably the most direct method of obtaining a majority-logic system.

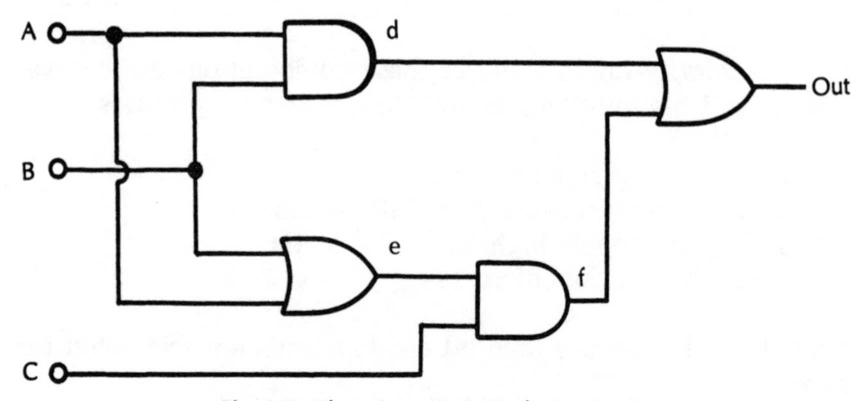

Fig. 8-5 Three-input majority-logic circuit.

To help you follow what is happening in this circuit, Table 8-5 shows a complete truth table. In addition to listing the input and output states, this truth table also shows the logic states at each of the intermediate points within the circuit. Clearly, as the number of inputs is increased, the majority-logic circuit inevitably becomes more complicated.

In this project you will build a five-input majority-logic demonstration circuit. I choose five inputs because three is too simple, but seven is quite complex. A five-input system seems like a good compromise.

In a five-input majority-logic system, at least three of the inputs are always at the same level, controlling the output state. The truth table for a five-input majority-logic circuit appears in Table 8-6.

Table 8-5 Complete Truth Table for Three-Input Majority-Logic-Demonstration Circuit.

Inputs						Output
A	B	C	d	e	f	0
0	0	0	0	0	0	0
0	0	1	0	1	0	0
0	1	0	0	1	0	
0	1	1	0	1	1	1
1	0	0	0	1	0	0
1	0	1	0	1	1	1
1	1	0	1	1	0	1
1	1	1	1	1	1	1

Table 8-6 Truth Table for Five-Input Majority-Logic Circuit.

Inputs					Output
A	B	C	D	E	
0	0	0	0	0	0
0	0	0	0	1	0
0	0	0	1	0	0
0	0	0	1	1	0
0	0	1	0	0	0
0	0	1	0	1	0
0	0	1	1	0	0
0	0	1	1	1	1
0	1	0	0	0	0
0	1	0	0	1	0
0	1	0	1	0	0
0	1	0	1	1	1
0	1	1	0	0	0
0	1	1	0	1	1
0	1	1	1	0	1
0	1	1	1	1	1
1	0	0	0	0	0
1	0	0	0	1	0
1	0	0	1	0	0
1	0	0	1	1	1
1	0	1	0	0	0
1	0	1	0	1	1
1	0	1	1	0	1
1	0	1	1	1	1
1	1	0	0	0	0
1	1	0	0	1	1
1	1	0	1	0	1
1	1	0	1	1	1
1	1	1	0	0	1
1	1	1	0	1	1
1	1	1	1	0	1
1	1	1	1	1	1

The five-input majority logic demonstration circuit appears in Fig. 8-6. Table 8-7 shows the parts list for this project. This circuit uses a total of six AND gates and five OR gates.

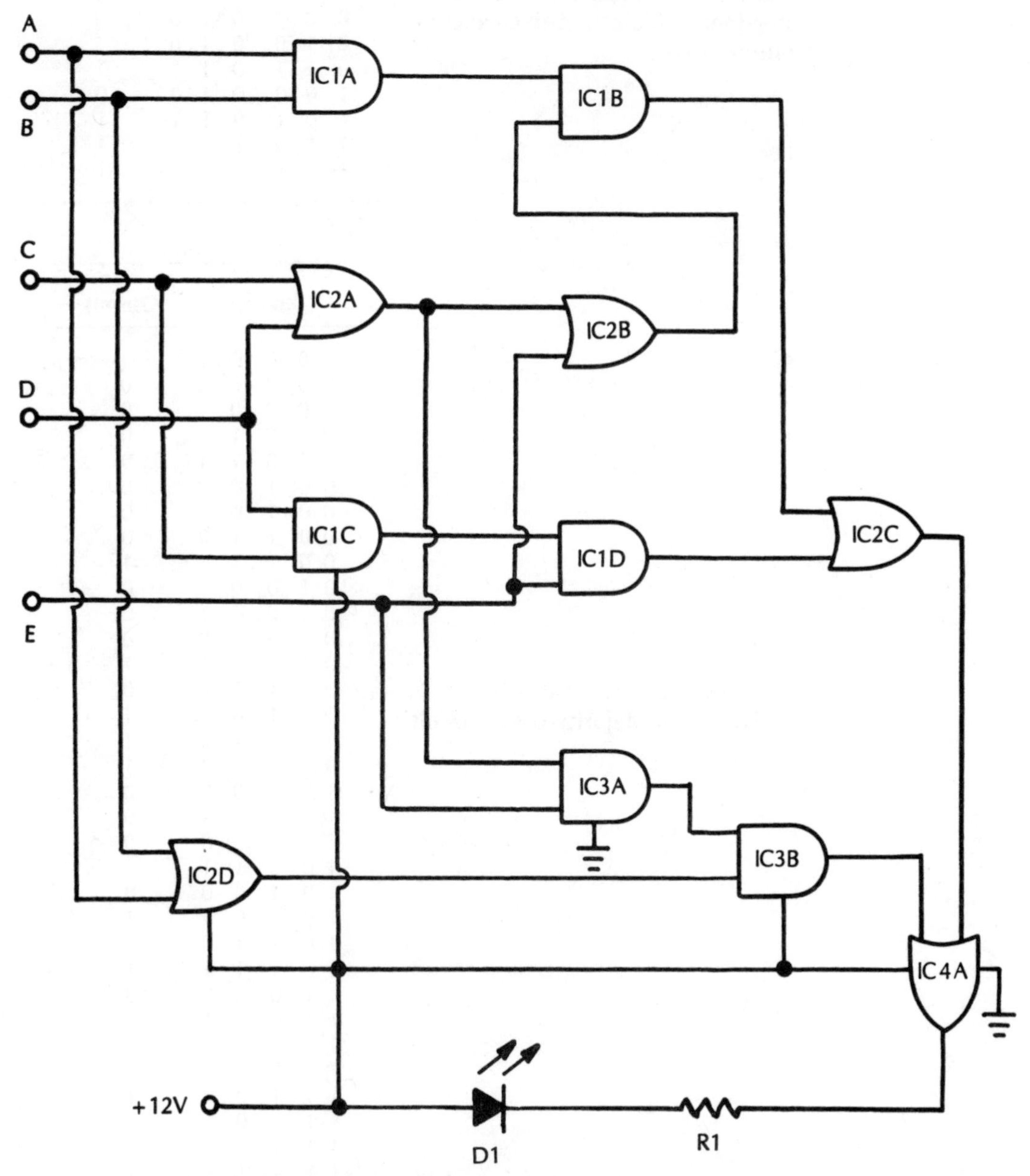

Fig. 8-6 *Project 47: Majority-Logic-Demonstration Circuit.*

Table 8-7 Parts List for the Majority-Logic-Demonstration Circuit of Project 47.

Part	Component
IC1, IC3	CD4081 quad AND gate
IC2, IC4	CD4071 quad OR gate
D1	LED
R1	330-ohm, 1/4-watt resistor

LED D1 indicates the output state. When the LED is lit, the output is a logic 1 (high). If the output is a logic 0 (low), the LED will be dark.

Resistor R1 limits the current through the LED. The exact value is not crucial. Anything in the 200 ohms to 1k range will do fine.

You can use the output from this majority-logic circuit to drive any digital input instead of (or perhaps in addition to) the LED if you so choose.

You can use any digital-input signals to drive the circuit. To manually demonstrate the majority-logic operation, just add a simple switch to each input. Figure 8-7 shows a suitable switching circuit. A simple SPST switch is used. When the switch is open (as shown), the appropriate gate input is grounded through resistor R, so it looks like a logic 0 (low). Closing the switch pulls the appropriate gate input up to V+, which looks like a logic 1 (high). Resistor R isolates the gate input from ground when the

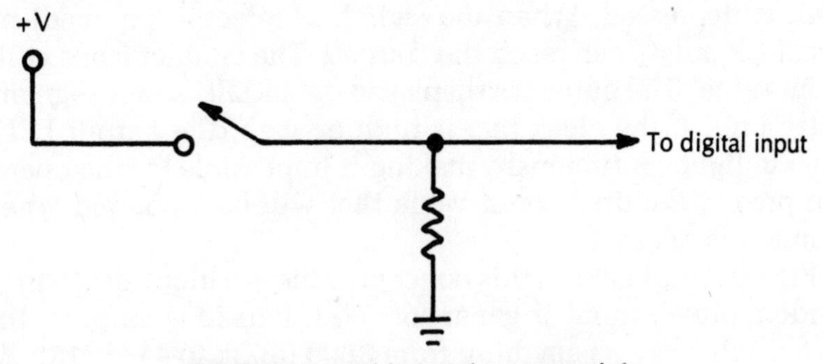

Fig. 8-7 A suitable switching network for the majority-logic-demonstration circuit of Fig. 8-6.

switch is closed. This resistor should have a fairly high value. I recommend something in the 1-megohm to 10-megohm range. The exact value is not important. You just need a large resistance here to prevent a short-circuit effect.

If you prefer, you could use the SPDT switch network illustrated in Fig. 8-8. No isolation resistor is needed here.

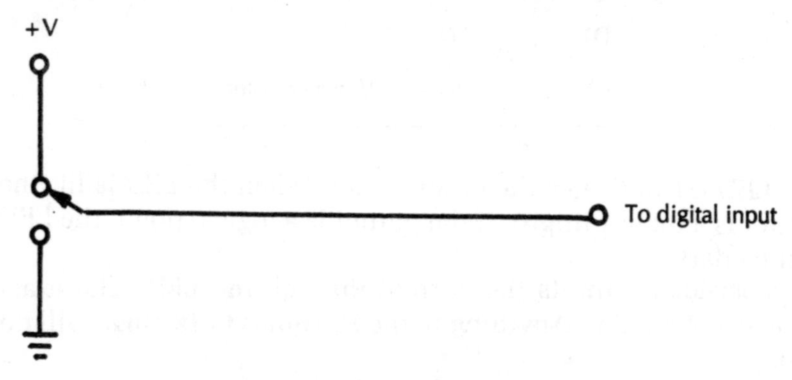

Fig. 8-8 *Alternate switching network for the majority-logic-demonstration circuit of Fig. 8-6.*

PROJECT 48: RANDOM-NUMBER GENERATOR

Occasionally, you might have a need for a digital circuit that can generate random (or quasi-random) numbers. Such a circuit can be useful for applications such as games, statistical experiments, "ESP testers," random-light displays, and random-music makers, among others.

Probably the easiest way to generate a random number in a digital circuit is to use a high-speed (high-frequency) clock signal that is fed to a counter while a momentary action pushbutton switch is depressed. When the switch is released (opened) no more clock pulses can reach the counter. The counter stops at its current value. The output is displayed on an LED seven-segment display unit. If the clock rate is high enough, the output LEDs appear to light continuously, making it impossible for the operator to predict the final count value that will be displayed when the button is released.

Figure 8-9 illustrates this concept. This is a block diagram of a random binary-number generator. With four LED outputs, the selected value can be anything from 0000 (0) up to 1111 (15). As

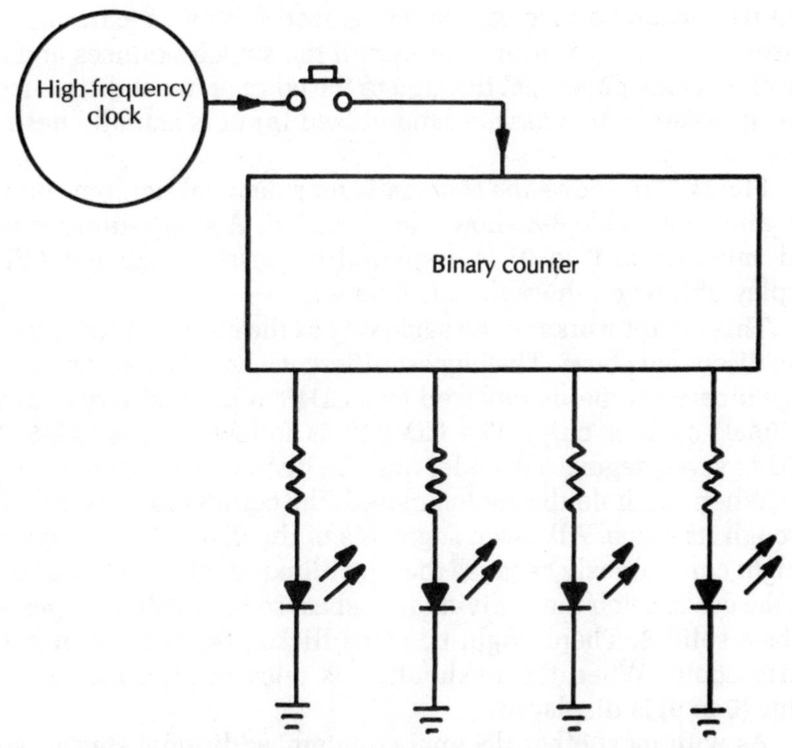

Fig. 8-9 Block diagram of a random binary-number generator.

you can see from the block diagram, this is a very simple and elegant approach to the problem of random-number generation.

An oscillator, or *astable multivibrator*, is used to generate the high-frequency clock signal. A 555 or 7555 timer is well suited to this type of application. The clock rate, of course, should be very high in frequency. I'd recommend a clock rate in the 10-kHz to 50-kHz range. The exact frequency is not particularly important, as long as it is too fast for the eye to follow, and the capabilities of the counter(s) are not exceeded.

The clock feeds (through the pushbutton) a four-stage binary counter made up of four flip-flops. When the pushbutton is held closed, the clock signal can get through to advance the counter. When the button is released, the counter stops incrementing, and the last count value is displayed on the output LEDs.

Incidentally, this is one type of digital circuit in which a switch debouncing network is never necessary. In most digital circuits, switch bounce can be recognized falsely as extra input pulses. In this application, however, if the switch bounces and a few more clock pulses get through to the counter, what difference does it make? In this case, a randomized input is actually desirable.

Figure 8-10 shows the schematic for a decimal random-number generator. Table 8-8 shows the parts list. A single-digit decimal value (from 0 to 9) is displayed on a seven-segment LED display unit when the switch is opened.

This circuit works in the same way as the simpler binary version discussed above. The biggest difference here is that the four-stage binary counter is replaced by a CD4518 BCD (binary coded decimal) counter chip. The CD4518 is followed by a CD4511 BCD-to-seven-segment decoder and the display unit itself.

When you hold the switch closed, the counter rapidly cycles through its range. All seven segments of the display unit appear to light continuously because they are blinking on and off too fast for the eye to catch the individual flashes, so the readout appears to be a solid 8. There might be some flicker, but it's nothing to worry about. When the pushbutton is released, the last count value (0 to 9) is displayed.

As with most other decimal counters, additional stages can be cascaded to allow for higher counts. Cascading two of the BCD counters allows for random numbers from 00 to 99.

There are two convenient ways to cascade the decimal-number generators. The approach shown in Fig. 8-11 is basically the same method used for a two or multidigit decimal counter. When counter A resets from 9 back to 0, it feeds a clock pulse into counter B, which counts the tens.

Alternately, you could use separate clocks to drive the two stages, as illustrated in Fig. 8-12. This gives you a more random effect, especially at lower clock frequencies. At higher clock frequencies, the difference between the two methods isn't very noticeable. The two clock oscillators should be set at different frequencies that are not related harmonically.

You might come up with a third approach on your own, but I'll forewarn you to avoid the setup shown in Fig. 8-13. This one might look like it should work, but it really doesn't. A single clock is used to drive the two counter stages simultaneously. This is a simple, economical, and direct method, but the results are very disappointing. If you think about it for a minute, you should

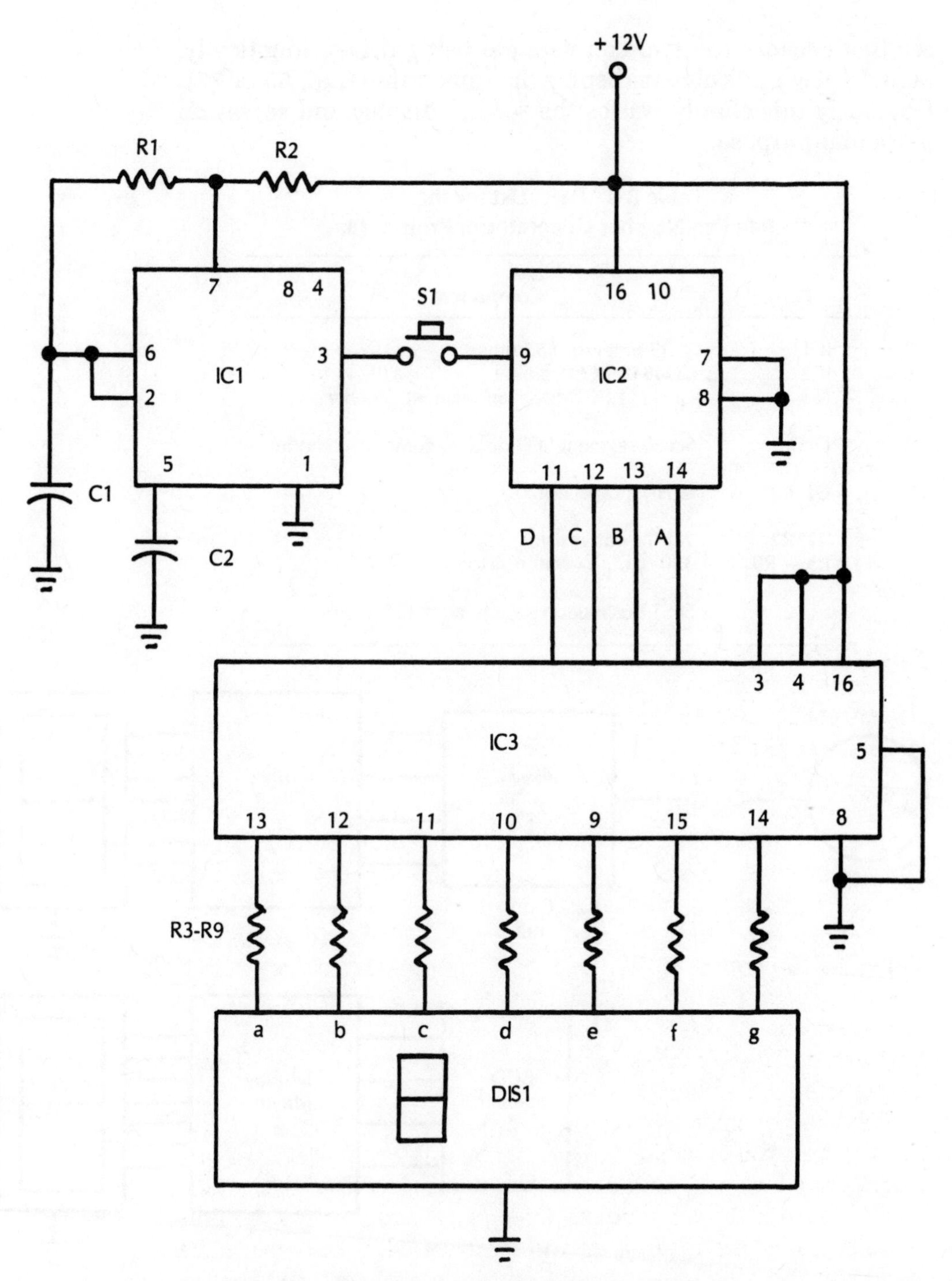

Fig. 8-10 Project 48: Random-Number Generator.

see that because the two counters are being driven identically, both display units always display the same value (i.e., 55 or 77). Obviously this simply wastes the second display and serves no particular purpose.

Table 8-8 Parts List for the Random-Number Generator of Project 48.

Part	Component
IC1	7555 timer (or 555 timer)
IC2	CD4518 BCD counter
IC3	CD4511 BCD-to-seven-segment decoder
DIS1	Seven-segment LED display, common-cathode
C1, C2	0.01-μF capacitor
R1, R2	1k, 1/4-watt resistor
R3 – R9	330-ohm, 1/4-watt resistor
S1	SPST pushbutton switch, normally open

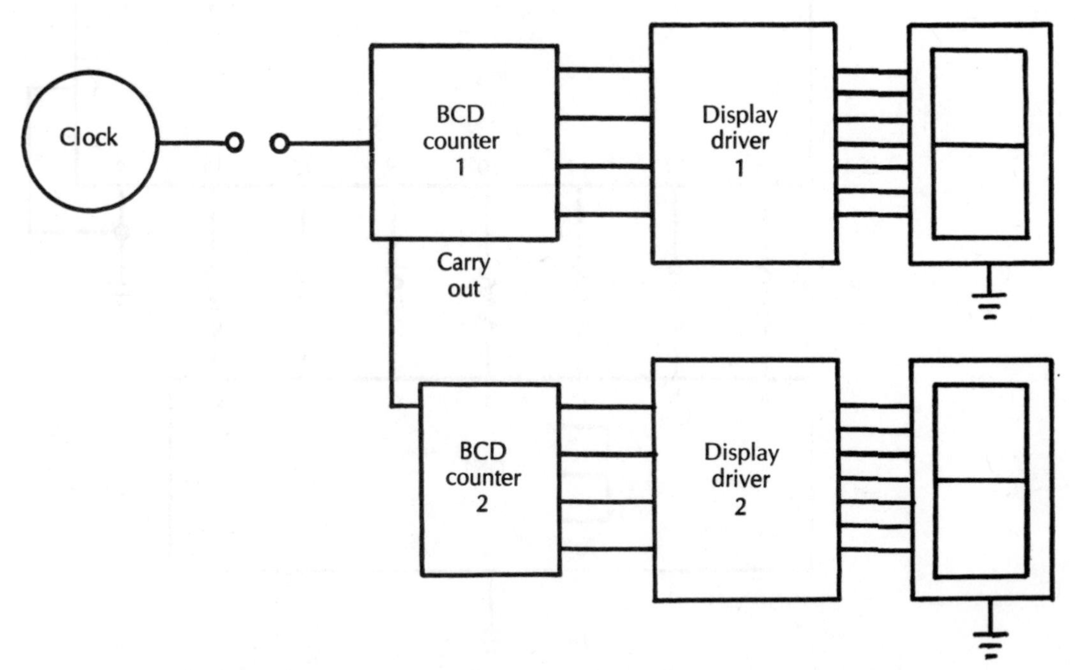

Fig. 8-11 Counter stages can be cascaded for two or more decimal digits.

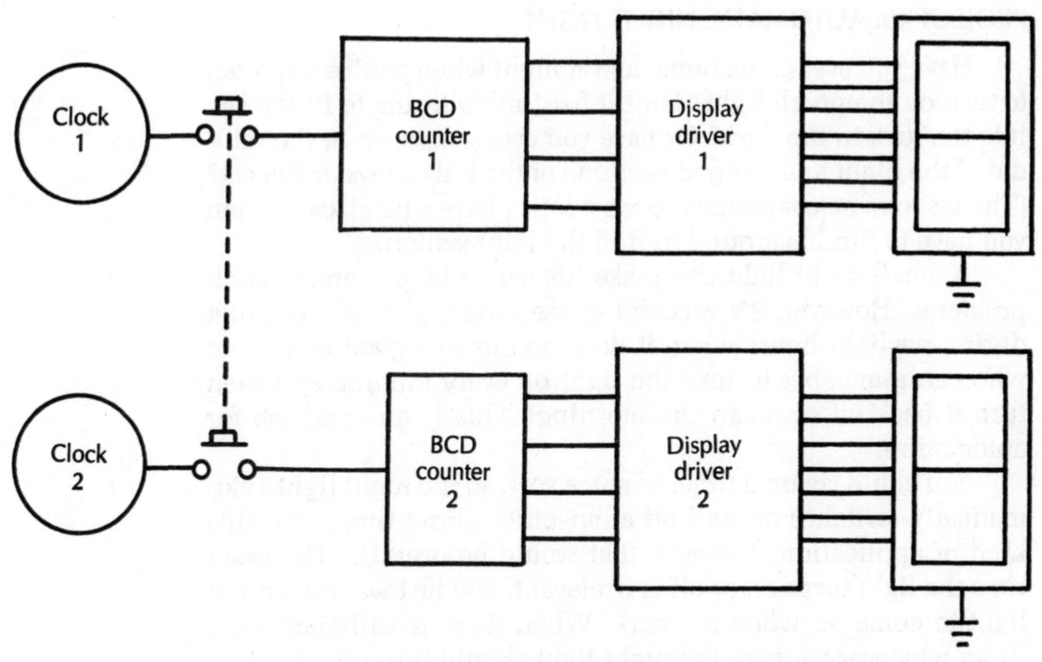

Fig. 8-12 *Separate clocks can be used to drive the two counter stages.*

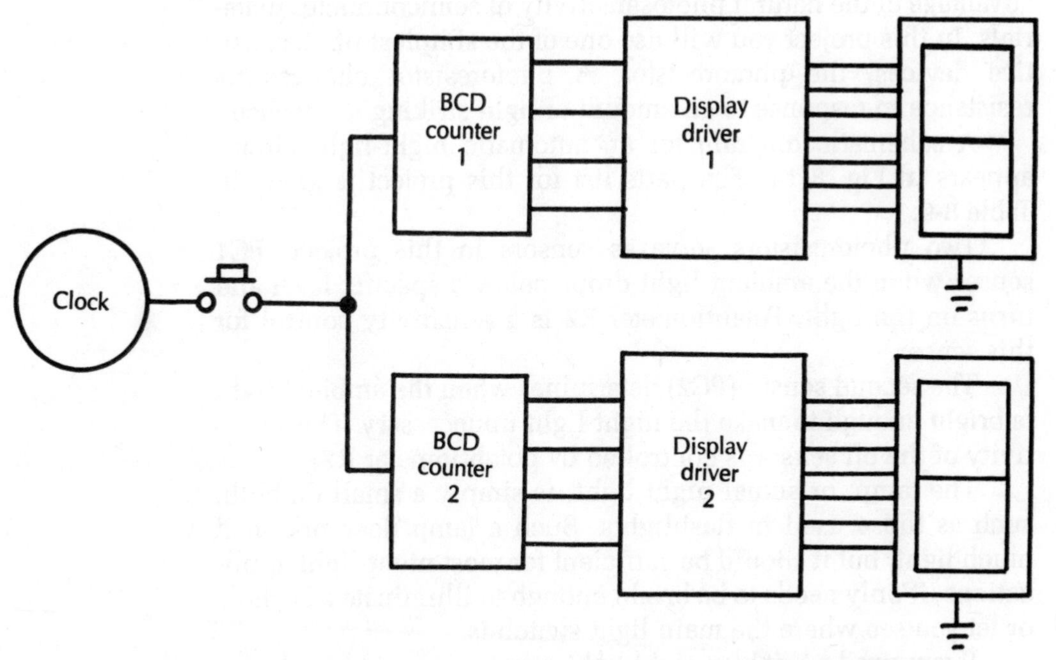

Fig. 8-13 *A single clock should not be used to drive both counter stages directly.*

PROJECT 49: AUTOMATIC NIGHT LIGHT

Have you ever come home late at night when you've forgotten to turn on the porch light? Isn't it frustrating trying to fit the key into the lock in the dark? Or have you ever gotten up in the middle of the night and tripped over one of the kids's toys or the cat? (The last one has happened to me a lot—I have a black cat.) Then you have to fumble around to find the light switch.

A small night light can make life easier by preventing such problems. However, it's wasteful to keep the night light burning during daylight hours when it does no one any good at all, but who can remember to turn the light on every evening and then turn it back off again in the morning? This is an ideal job for automation.

You could rig up a timer of some sort, so the night light automatically switches on and off at specific, preset times. For this kind of application, however, that would be overkill. The exact time the light turns on or off is irrelevant. You just want the night light to come on when it's dark. When there is sufficient light (from whatever source), the night light should turn off.

A number of electronic components are designed to take advantage of the natural photosensitivity of semiconductor materials. In this project you will use one of the simplest photosensitive devices, the photoresistor. A *photoresistor* changes its resistance in response to an amount of light striking its surface.

A schematic diagram for an automatic night-light circuit appears in Fig. 8-14. The parts list for this project is given in Table 8-9.

Two photoresistors serve as sensors in this project. PC1 senses when the ambient light drops below a specific level and turns on the light. Potentiometer R2 is a sensitivity control for this sensor.

The second sensor (PC2) determines when the ambient light is bright enough to make the night light unnecessary. The sensitivity of the off sensor is controlled by potentiometer R7.

The lamp, or actual night light, is simply a small dc bulb, such as those used in flashlights. Such a lamp does not shed much light, but it should be sufficient for most night-light applications. It only needs to be bright enough to illuminate a keyhole or let you see where the main light switch is.

If you need a brighter night light, you can use a larger lamp.

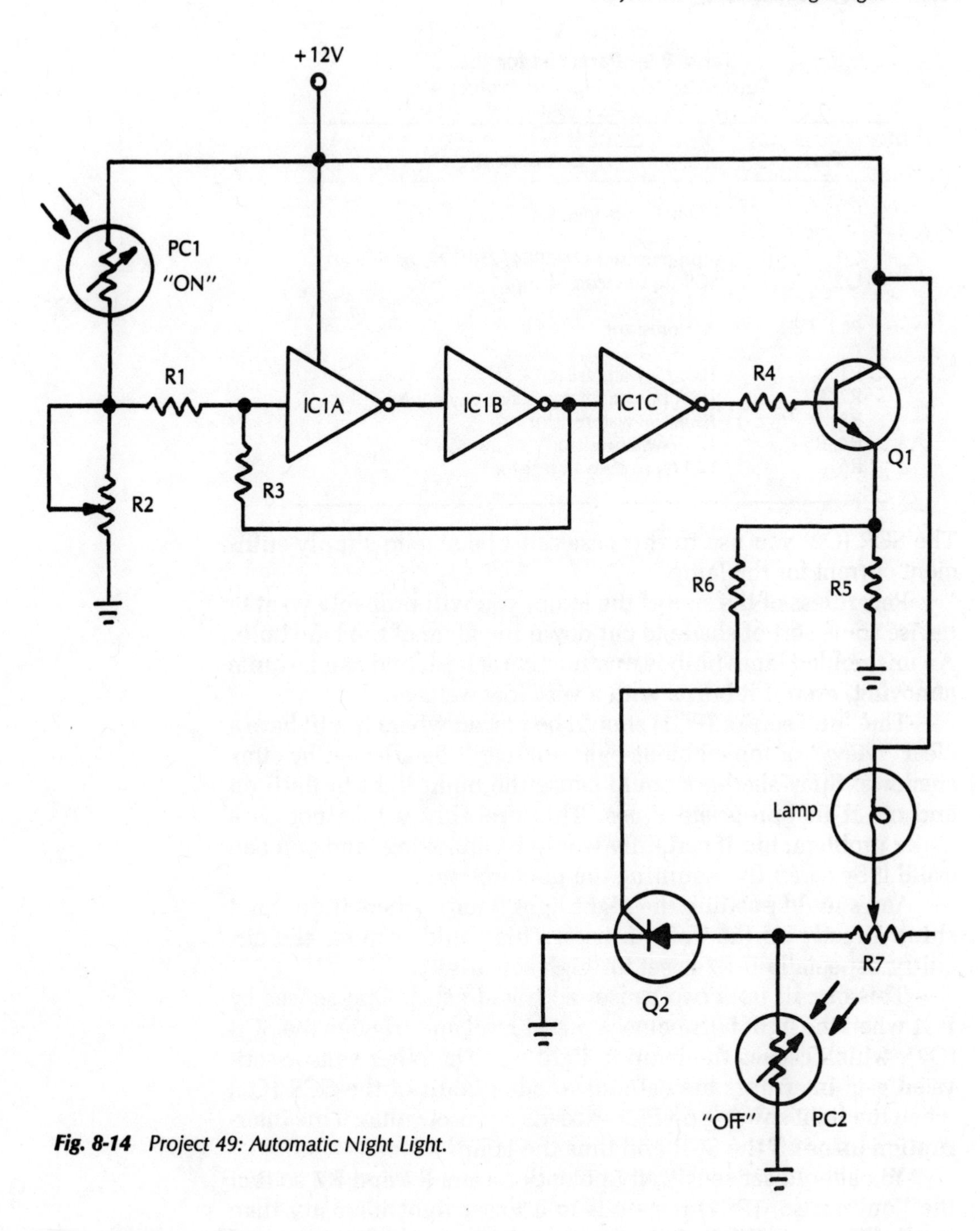

Fig. 8-14 Project 49: Automatic Night Light.

Table 8-9 Parts List for the Automatic Night Light of Project 49.

Part	Component
IC1	CD4049 hex inverter
Q1	Npn transistor (2N3904, 2N2222, or similar)
Q2	SCR (to suit load—lamp)
PC1, PC2	Photoresistor
R1	10k, 1/4-watt resistor
R2, R7	100k potentiometer (sensitivity adjust)
R3	100k, 1/4-watt resistor
R4, R5	1k, 1/4-watt resistor
R6	120-ohm, 1/2-watt resistor

The SCR (Q2) you use in this case must be able to supply sufficient current for the lamp.

Regardless of the size of the lamp, you will probably want to devise some sort of shade to cut down the glare of the bare bulb. An unshielded lamp bulb is unattractive at best, and can be quite annoying, even if it burns with a very low wattage.

The "on" sensor (PC1) should be placed where it will have a clear "view" of the ambient light and can't be affected by stray shadows. Stray shadows could cause the night light to flash on and off at inappropriate times. This probably would not be a major problem, but it certainly would be annoying, and you can avoid it by carefully mounting the photoresistor.

You should position the night light (lamp) where it does not shine directly on the "off" sensor. This could confuse the circuitry, especially if R7 is set for high sensitivity.

This circuit uses two sensor-activated gates. One senses by PC1 when the light falls below a preset level and triggers the SCR (Q2), which causes the lamp to light up. The other sensor-activated gate interrupts the cathode/anode circuit of the SCR (Q2) when the light shining on PC2 exceeds a preset value. This interruption turns off the SCR and thus the lamp.

You should set sensitivity potentiometers R2 and R7 so that the "on" sensor (PC1) responds to a lower light intensity than the "off" sensor (PC2). Otherwise the lamp might blink on and off under some conditions.

I certainly hope you enjoy these projects as much as I did. Feel free to experiment with and modify any or all of the circuits to suit your own individual needs. Customizing projects is one of the most exciting aspects of an electronics hobby.

Index